D1739544

Number
Fluency

Peter Clarke

*Developing mental fluency
in numerical skills*

Year 2

Acknowledgements

The author wishes to thank Brian Molyneaux and Mike Askew
for their valuable contribution to this publication.

Published by BEAM at Nelson Thornes
Delta Place
27 Bath Road
Cheltenham GL53 7TH

Telephone 01242 267 287
Fax 01242 253 695
Email cservices@nelsonthornes.com

ISBN 978 1 9062 2481 3

British Library Cataloguing-in-Publication Data
Data available

Cover photo: Dreamstime

Printed in Croatia by Zrinski

14 13 12 11 10 / 9 8 7 6 5 4 3 2 1

Contents

Introduction

The BEAM *Number Fluency* series is a set of six books, one for each year from Year 1 to Year 6, that aims to support the development of number fluency in basic numerical skills through individual and paired activities.

The books comprise teaching strategies plus photocopiable activity resources that, for ease of teaching and photocopy reproduction, are also available online at no extra cost to book purchasers (see page 9).

Each book covers key learning objectives based on the Primary Numeracy Strategy (PNS): *Primary Framework for Mathematics* (2006). These objectives are organised into six sections, each of which addresses a key aspect of becoming fluent in number.

The six sections are the same across all six books and are:

1. Comparing and ordering numbers
2. Place value and partitioning
3. Understanding and using tens
4. Deriving and recalling addition and subtraction facts, and using that knowledge
5. Deriving and recalling multiplication and division facts, and using that knowledge
6. Mental calculation methods

The chart on page 10 links each of these six sections to the relevant Year 2 strand objective and planning blocks and units from the PNS *Primary Framework for Mathematics* (2006). Refer to this chart when choosing a section.

Children develop fluency in number through a combination of four key elements:

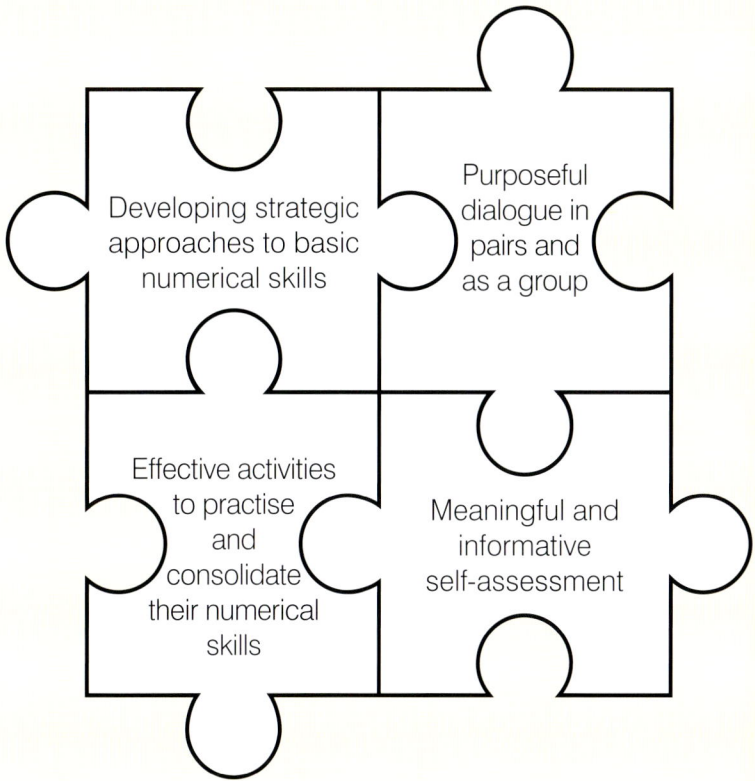

Developing strategic approaches to basic numerical skills

Purposeful dialogue in pairs and as a group

Effective activities to practise and consolidate their numerical skills

Meaningful and informative self-assessment

 ## Content

Each section begins with an introduction that includes:

- strategic approaches to develop fluency skills
- a brief description of the individual and paired activities
- teaching suggestions for ways to help children develop the fluency skills and provide further practice
- an individual child self-assessment record

You will find a more detailed explanation of each of these features on pages 8 and 9.

Each of the six sections is divided into two levels: Level 1 (easier) and Level 2 (harder). There are two types of activities at each level: an individual activity and a paired activity. Both require children to work in pairs.

The individual activity contains two worksheets: A and B. Child A uses worksheet A; Child B uses worksheet B. The children work individually, but each worksheet contains the answers to their partner's questions. This enables each child to correct their partner's answers and to discuss the results.

For each of the paired activities, there are two worksheets: A and B. Child A uses worksheet A; Child B uses worksheet B. Here, the children need to work together to complete their worksheets. These activities are either self-checking or pair-checking.

The diagram below aims to explain the structure of *Number Fluency*, using Year 2 Section 1 as an example.

Number Fluency **Year 2**

Section 1

Comparing and ordering numbers

PNS *Framework for Mathematics* (2006) objective:

- Order two-digit numbers and position them on a number line

Number Fluency Level 1 objective:

- Compare and order two-digit numbers and position them on a partially numbered number line [lots of numbers already in place]

Number Fluency Level 2 objective:

- Compare and order two-digit numbers and position them on a partially numbered number line [fewer numbers already in place]

| Individual activity 1A | Paired activity 1A |
| Individual activity 1B | Paired activity 1B |

| Individual activity 2A | Paired activity 2A |
| Individual activity 2B | Paired activity 2B |

The chart on page 11 shows Level 1 and Level 2 objective coverage for each of the six sections in *Number Fluency* Year 2. Refer to this chart to differentiate not only for particular individuals and pairs of children, but also when choosing the level of work that is most suitable for a specific class, as well as the time of year in which you are teaching or consolidating the objective.

Suggestions for using *Number Fluency*

Number Fluency is a flexible resource that you can use in many different ways. One suggestion includes:

> Decide which objective you wish to develop fluency in (refer to the appropriate section).

↓

> Choose the appropriate level (Level 1 or 2).

↓

> Provide each pair with the individual activity to gain some self-assessed awareness of their level of fluency (Child A using worksheet A and Child B using worksheet B).

↓

> Children work together on the paired activity (A and B) with the explicit intention of supporting each other to improve their fluency.

↓

> Children repeat the individual activity to check their progress (this time, Child A using worksheet B and Child B using worksheet A).

As well as the individual activities and paired activities, *Number Fluency* offers:

- strategic approaches to develop fluency
- further activities to develop fluency

Use these suggestions as you see fit:

- before children complete the first individual activity
- after children complete the first individual activity and before they do the paired activity
- after children complete the paired activity and before they do the second individual activity
- after children complete the second individual activity

Number Fluency also provides an individual child's self-assessment record for each section. Encourage the children to use these records to monitor their own progress and to see their improvement against their own baseline as opposed to being compared against other class members.

 ## The importance of promoting effective speaking and listening in developing children's mathematical understanding

Number Fluency recognises the importance of getting children to work collaboratively. By working as a pair, children learn from each other, confirming their mathematical knowledge and identifying for themselves, in a non-threatening environment, any misconceptions they may hold.

The recommendations in both the *Independent Review of Mathematics Teaching in Early Years Settings and Primary Schools* (2008) and the *Independent Review of the Primary Curriculum* (2008) reported on the importance of actively promoting speaking and listening in mathematics.

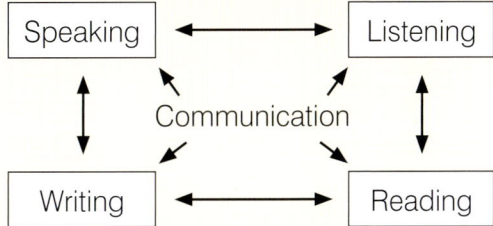

Each of the planning blocks and units in the PNS *Primary Framework for Mathematics* (2006) shows links with the relevant objectives from PNS *Speaking, Listening, Leaning: Working with children in Key Stages 1 and 2* (2003). All of the activities in *Number Fluency* Year 2 aim to cover each of the following Speaking, Listening, Learning Year 2 objectives:

Speaking

- Speak with clarity and use intonation when reading and reciting texts

- Use language and gesture to support the use of models/diagrams/displays when explaining

- Tell real or imagined stories (using conventions of familiar story language)

Listening

- Listen to others in class, ask relevant questions and follow instructions

- Respond to presentations by describing characters, repeating some highlights and commenting constructively

- Listen to a talk by an adult, remember some specific points and identify what they have learned

Group discussion and interaction

- Listen to each other's views and preferences, agree the next steps to take and identify contributions by each group member

- Ensure everyone contributes, allocate tasks, consider alternatives and reach agreement

- Work effectively in groups by ensuring each group member takes a turn, challenging, supporting and moving on

- Explain their views to others in a small group and decide how to report the group's views to the class

How to use this book

Each of the six sections in *Number Fluency* Year 2 includes two pages offering advice on how to teach the objective strategically and how the individual and paired activities support this.

Strategic approaches to developing fluency

Offers several approaches as to how to teach the objective strategically. Use these activities as part of the main teaching activity before children complete an individual or paired activity.

Individual and paired activities

Description of each of the individual and paired activities explaining the purpose of the activity, and one or two questions that you might use to discuss the activity with the children

Level 1
Individual activity

Level 1
Paired activity

Level 2
Individual activity

Level 2
Paired activity

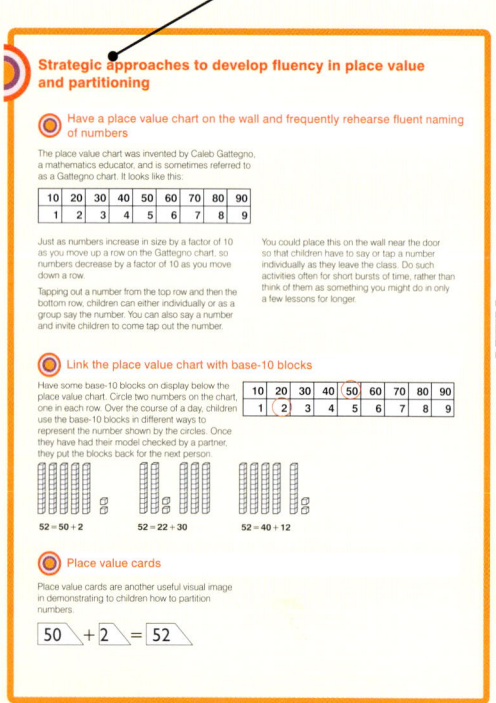

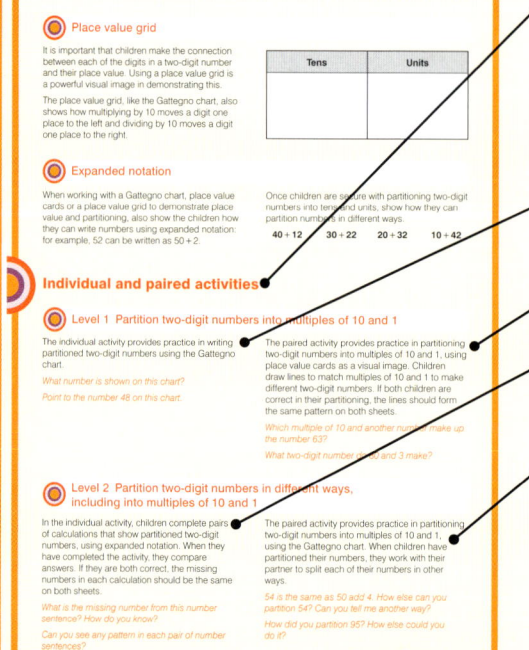

Each of the six sections in *Number Fluency* Year 2 also includes further activities to develop fluency and an individual child record sheet.

Further activities to develop fluency

Suggestion of further teaching activities to develop the fluency further. Use these activities either before or after the children complete the individual and paired activities.

My record sheet

Each section includes an individual child self-assessment sheet.

The top half of the sheet states the fluency objectives in child speak with 'smileys' (☺☺☺) to record self-assessment before and after the activities.

The bottom half of the sheet has three or four open response questions to encourage the child to reflect on their learning.

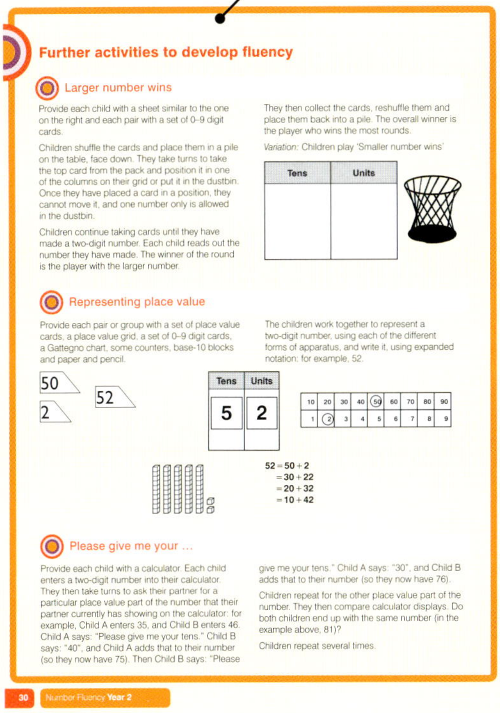

Further activities to develop fluency

Larger number wins

Provide each child with a sheet similar to the one on the right and each pair with a set of 0–9 digit cards.

Children shuffle the cards and place them in a pile on the table, face down. They take turns to take the top card from the pack and position it in one of the columns on their grid or put it in the dustbin. Once they have placed a card in a position, they cannot move it, and one number only is allowed in the dustbin.

Children continue taking cards until they have made a two-digit number. Each child reads out the number they have made. The winner of the round is the player with the larger number.

They then collect the cards, reshuffle them and place them back into a pile. The overall winner is the player who wins the most rounds.

Variation: Children play 'Smaller number wins'.

Tens	Units

Representing place value

Provide each pair or group with a set of place value cards, a place value grid, a set of 0–9 digit cards, a Gattegno chart, some counters, base-10 blocks and paper and pencil.

The children work together to represent a two-digit number, using each of the different forms of apparatus, and write it, using expanded notation: for example, 52.

Tens	Units
5	2

$$52 = 50 + 2$$
$$= 30 + 22$$
$$= 20 + 32$$
$$= 10 + 42$$

Please give me your …

Provide each child with a calculator. Each child enters a two-digit number into their calculator. They then take turns to ask their partner for a particular place value part of the number that their partner currently has showing on the calculator: for example, Child A enters 35, and Child B enters 46. Child A says: "Please give me your tens." Child B says: "40", and Child A adds that to their number (so they now have 75). Then Child B says: "Please

give me your tens." Child A says: "30", and Child B adds that to their number (so they now have 76).

Children repeat for the other place value part of the number. They then compare calculator displays. Do both children end up with the same number (in the example above, 81)?

Children repeat several times.

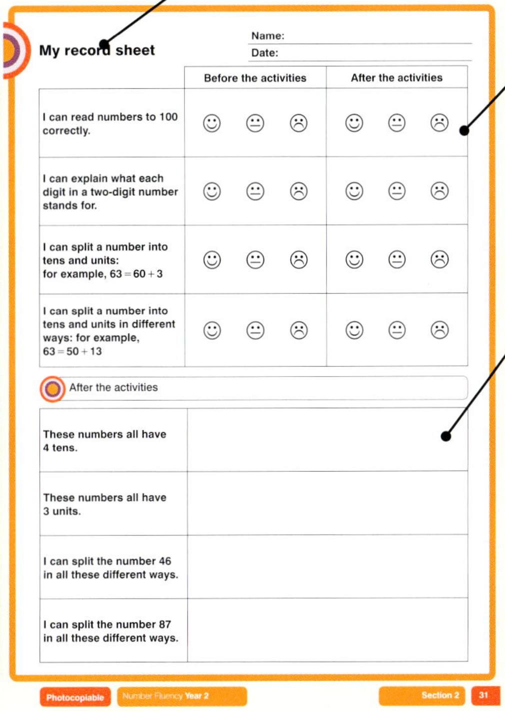

My record sheet

Name: _____
Date: _____

	Before the activities			After the activities		
I can read numbers to 100 correctly.	☺	☺	☹	☺	☺	☹
I can explain what each digit in a two-digit number stands for.	☺	☺	☹	☺	☺	☹
I can split a number into tens and units: for example, 63 = 60 + 3	☺	☺	☹	☺	☺	☹
I can split a number into tens and units in different ways: for example, 63 = 50 + 13	☺	☺	☹	☺	☺	☹

After the activities

These numbers all have 4 tens.	
These numbers all have 3 units.	
I can split the number 46 in all these different ways.	
I can split the number 87 in all these different ways.	

Online access

Purchasers of the book can access the Record sheets for each section of each book, and all of the photocopiable Activity pages, online, for ease of photocopying or for interactive whiteboard display. To use these go to the following unique web address: www.beam.co.uk/numberfluency-J35

Chart linking to the PNS *Primary Framework for Mathematics* (2006)

Number Fluency section	PNS *Primary Framework for Mathematics* (2006)		
	Strand	Objective	Planning block and unit
1. Comparing and ordering numbers	2: Counting and understanding number	Order two-digit numbers and position them on a number line	A1, A3
2. Place value and partitioning	2: Counting and understanding number	**Explain what each digit in a two-digit number represents, including numbers where 0 is a place holder; partition two-digit numbers in different ways, including into multiples of 10 and 1**	A1, A2, A3
3. Understanding and using tens	4: Calculating	**Add mentally a one-digit number or a multiple of 10 to any two-digit number**	A1, D1, A2, D2, A3, D3
4. Deriving and recalling addition and subtraction facts, and using that knowledge	3: Knowing and using number facts	**Derive and recall all addition and subtraction facts for each number to at least 10 and all pairs of multiples of 10 with totals to 100**	B1, B2, B3
5. Deriving and recalling multiplication and division facts, and using that knowledge	3: Knowing and using number facts	Derive and recall multiplication facts for the 2, 5 and 10 times tables and the related division facts	B1, E1, B2, E2, B3, E3
6. Mental calculation methods	4: Calculating	**Use the symbols $+$, $-$, $\times$, $\div$ and $=$ to record and interpret number sentences involving all four operations; calculate the value of an unknown in a number sentence: for example, $\square \div 2 = 6$, $30 - \square = 24$**	E1, A2, E2, A3, E3

Note: Key objectives are in **bold**.

Individual and paired activities

Number Fluency section	Level 1 Objective coverage	Pages Individual activity 1A, 1B	Pages Paired activity 1A, 1B	Level 2 Objective coverage	Pages Individual activity 2A, 2B	Pages Paired activity 2A, 2B
1. Comparing and ordering numbers	Compare and order two-digit numbers and position them on a partially numbered number line [lots of numbers already in place]	18, 19	20, 21	Compare and order two-digit numbers and position them on a partially numbered number line [fewer numbers already in place]	22, 23	24, 25
2. Place value and partitioning	Partition two-digit numbers into multiples of 10 and 1	32, 33	34, 35	Partition two-digit numbers in different ways, including into multiples of 10 and 1	36, 37	38, 39
3. Understanding and using tens	Add mentally a one-digit number to any two-digit number	46, 47	48, 49	Add mentally a one-digit number to any two-digit number and a multiple of 10 to any two-digit number	50, 51	52, 53
4. Deriving and recalling addition and subtraction facts, and using that knowledge	Derive and recall all addition facts for each number to at least 10	60, 61	62, 63	Derive and recall all pairs of multiples of 10 with totals to 100	64, 65	66, 67
5. Deriving and recalling multiplication and division facts, and using that knowledge	Derive and recall multiplication facts for the 2, 5 and 10 times tables	74, 75	76, 77	Derive and recall division facts corresponding to the 2, 5 and 10 times tables	78, 79	80, 81
6. Mental calculation methods	Calculate the value of an unknown in a number sentence [for example, $30 - \square = 24$], using the symbols $+$, $-$ and $=$	88, 89	90, 91	Calculate the value of an unknown in a number sentence [for example, $\square \div 2 = 6$], using the symbols $\times$, $\div$ and $=$	92, 93	94, 95

Comparing and ordering numbers

Level 1
• Compare and order two-digit numbers and position them on a partially numbered number line [lots of numbers already in place]

Level 2
• Compare and order two-digit numbers and position them on a partially numbered number line [fewer numbers already in place]

Strategic approaches to develop fluency in comparing and ordering numbers

Work on placing numbers on number lines

There is a lot of research in psychology that shows that the brain stores numbers in a linear form. Putting numbers on an empty number line and getting them well spaced is a key skill in understanding the order of numbers.

Children need to have experience of ordering numbers on both partially numbered and unnumbered number lines, as well as an empty number line.

Partially numbered number line

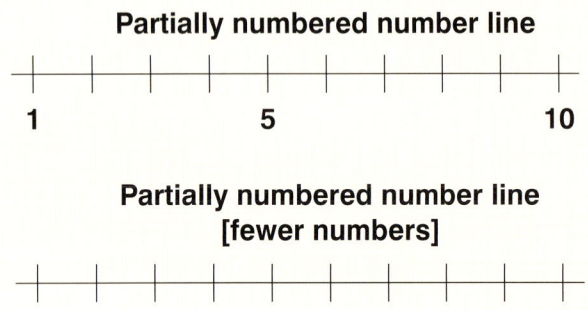

Partially numbered number line [fewer numbers]

Unnumbered number line

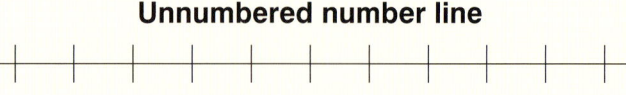

Put two numbers on the board: for example, 1 and 50. Use these to mark the ends of an empty number line:

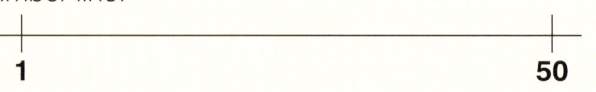

Approximately, mark the centre of the line. Children agree in pairs what would be a suitable number to write there. Discuss and agree on an answer as a class.

Mark other points on the line and reach a consensus on sensible numbers to write there.

Also provide examples where the number range does not always start with 0 or 1:

Emphasise how the language of numbers helps

Write down two two-digit numbers on two pieces of paper: for example, 87 and 43. Invite two children to help and give one number to each of them; the class should not know what the numbers are.

Each child takes it in turn to start to read out their number, but they are only going to read one digit's

value at a time. So child A says: "Eight tens", and Child B says: "Four tens." Can the class decide who has the larger number without hearing any more?

Repeat with numbers like 71 and 17 or 83 and 43.

Model numbers with base-10 materials

Help children get a feel for numbers by putting out a collection of base-10 materials. Put the bricks for 46, for example, on a paper plate or a piece of card and cover them with a cloth. Start with the bricks set out in the their TU columns. Tell the children that they are not going to have time to count the bricks as you are only going to allow them a quick glimpse. Quickly remove the cloth,

then cover up the bricks again. What is the range of estimates that the children come up with? What strategies did they use to come up with numbers so quickly?

Move on to the bricks being displayed irregularly. How good can the children get at making 'quick glimpse' estimations?

In their TU columns

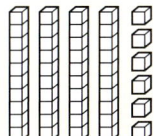

Irregularly displayed

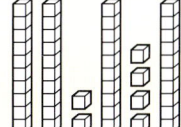

 ## Move between representations

Children need to be fluent in moving between four representations of numbers: spoken, recorded in numerals, positioned on a number line and modelled with base-10 blocks.

Regularly start with one of these representations and encourage the children to set up the others. Which do they find easiest to start with? Which is the most challenging?

Individual and paired activities

 ### Level 1 Compare and order two-digit numbers and position them on a partially numbered number line [lots of numbers already in place]

The individual activity provides practice in selecting the largest and smallest number from a collection of numbers and then positioning them on a number line. When children have completed the work individually, focus their attention on the number lines.

[Covering one of the numbers with your finger] Which number have I covered up?

What would be the number before the first number on the line?

The paired activity provides an opportunity to consolidate comparing two numbers and to practise reading out numbers to 100. When they have completed the activity, both children's number lines should be identical with the same numbers missing from each of their number lines.

Do both of you have the same numbers written on your number lines? Why/Why not?

What numbers are missing from this number line?

 ### Level 2 Compare and order two-digit numbers and position them on a partially numbered number line [fewer numbers already in place]

The individual activity provides an opportunity for children to order sets of five numbers from smallest to largest. Children then place two numbers from each set on partially numbered number lines. When they have completed the activity, both children's number lines should be identical, with the same numbers missing from their number lines.

[Pointing to one of the unnumbered numbers on a number line] Which number belongs here? How do you know?

Look at the 12 numbers in the circles. Which is the largest of these numbers? Which is the smallest?

The paired activity provides an opportunity to consolidate comparing and ordering numbers by finding the smallest and largest numbers from sets of numbers and to practise reading and writing numbers to 100. When they have completed the activity, both children's number lines should be identical with the same numbers missing from each of their number lines.

Which is the largest of these numbers? How do you know? Which is the smallest?

Point to and say some of the missing numbers from the number lines.

Further activities to develop fluency

 ## Lay down your cards

Provide each group with a pile of unordered two-digit number cards. Each child takes five cards. Children lay their cards on the table in front of them, face up. The child who has the card nearest to 50 starts and places the card in the centre of the table: for example, 56. Each child then looks at their cards, and the two children who have the next smallest and largest numbers place their cards beside the initial number card, either to the left (if the number is smaller) or to the right (if the number is larger: for example, 53 56 62. The winner is the first player to lay all their five cards in the row.

 ## Still in order

Write the following on the board:

13 □ 27 □ 38 □ 42

Can the children suggest numbers for each box so that all the numbers are in order?

What other numbers could go in the boxes?

 ## Identifying numbers on a number line

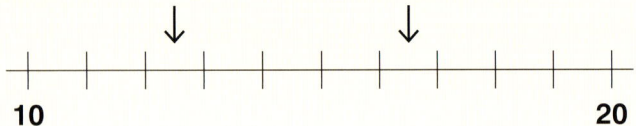

Draw the number line on the left on the board.

Can the children say which numbers belong on the unmarked intervals between the arrows?

 ## 100-grid jigsaw

Cut up a 100-grid into small pieces.

Children lay out all the small pieces on the table, face up.

How quickly can they reassemble the 100-grid?

1	2	3	4	5	6	7	8	9	10
11	12	13	14	15	16	17	18	19	20
21	22	23	24	25	26	27	28	29	30
31	32	33	34	35	36	37	38	39	40
41	42	43	44	45	46	47	48	49	50
51	52	53	54	55	56	57	58	59	60
61	62	63	64	65	66	67	68	69	70
71	72	73	74	75	76	77	78	79	80
81	82	83	84	85	86	87	88	89	90
91	92	93	94	95	96	97	98	99	100

 ## Make and order

Provide each child with three digit cards: for example, 3, 6, 8. Children place any two cards together to make a two-digit number such as 36. They investigate making different two-digit numbers (six different two-digit numbers are possible), then write them in order.

My record sheet

Name:

Date:

	Before the activities			After the activities		
I can compare two numbers and say which is smaller or larger.	☺	😐	☹	☺	😐	☹
I can order a set of numbers to 100.	☺	😐	☹	☺	😐	☹
I can place a set of numbers to 100 on a number line.	☺	😐	☹	☺	😐	☹

◉ After the activities

Here are some numbers in order, smallest to largest.

Here are some numbers to 100 written on a number line.

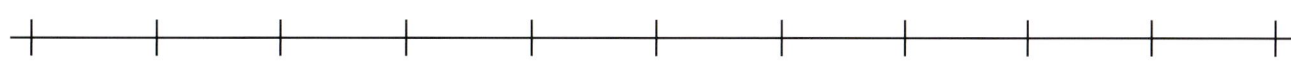

Here are some more numbers to 100 written on a number line.

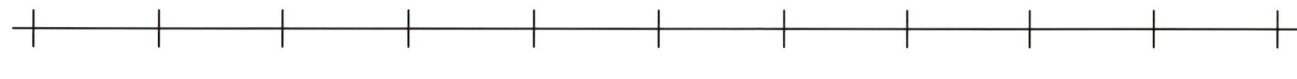

Individual activity 1A

Name: _____

Date: _____

Compare and order two-digit numbers and position them on a partially numbered number line [lots of numbers already in place]

For each set of numbers:
- draw a circle ◯ around the largest number
- draw a box ▢ around the smallest number

| 37 | 26 | 28 | 55 | 51 |

| 53 | 58 | 92 | 88 | 91 |

| 59 | 61 | 60 | 26 | 24 |

| 90 | 24 | 22 | 28 | 74 |

| 81 | 21 | 87 | 19 | 77 |

| 93 | 94 | 59 | 58 | 90 |

Look at the numbers you circled ◯ or drew a box ▢ around.
Write these numbers on the number lines.

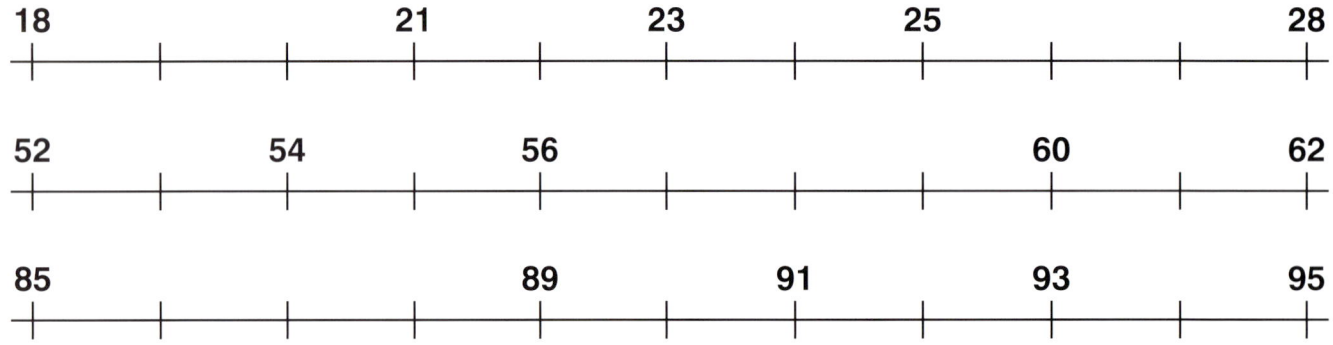

Check your answers with your partner.

Talk with your partner about how you knew where to write the numbers on the number line.

Answers to 1B

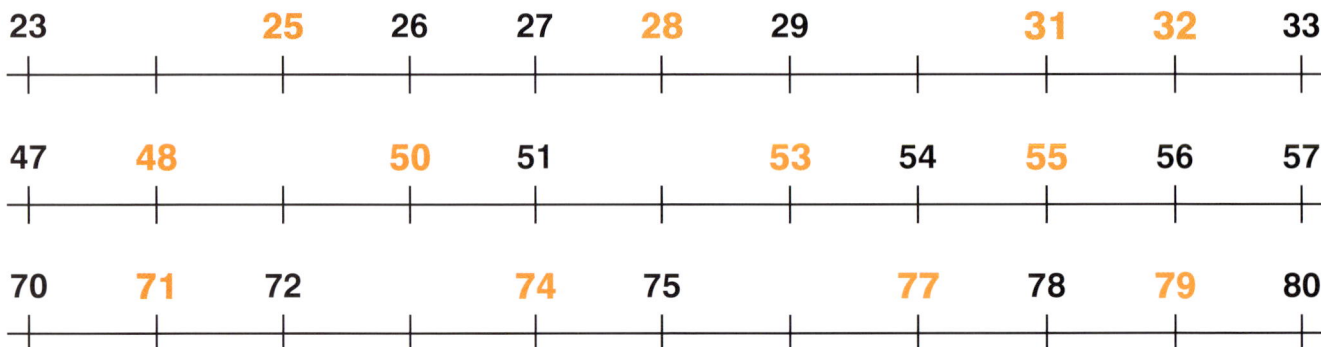

Name:

Date:

Compare and order two-digit numbers and position them on a partially numbered number line [lots of numbers already in place]

For each set of numbers:
- draw a circle ◯ around the largest number
- draw a box ☐ around the smallest number

61	28	32	71	70

38	32	35	53	51

27	75	25	77	72

74	31	73	71	54

60	70	79	50	55

54	55	49	51	48

Look at the numbers you circled ◯ or drew a box ☐ around.
Write these numbers on the number lines.

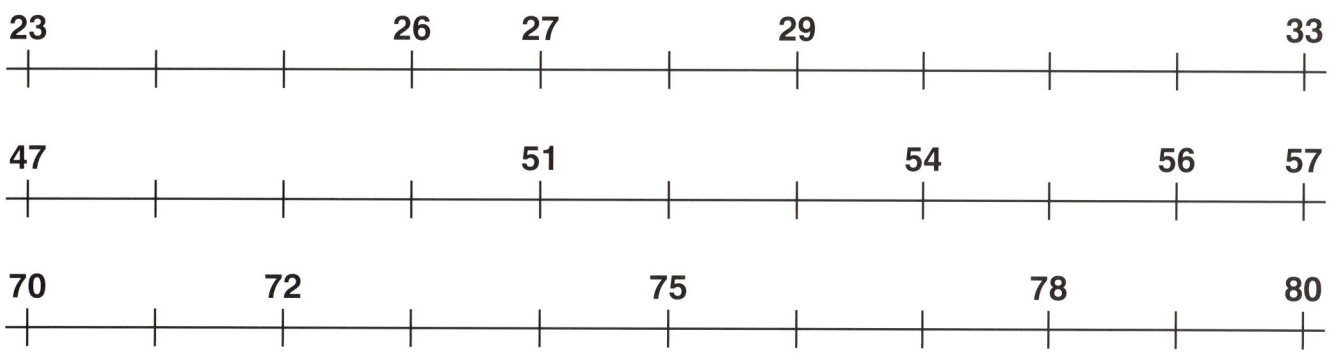

Check your answers with your partner.

Talk with your partner about how you knew where to write the numbers on the number line.

Answers to 1A

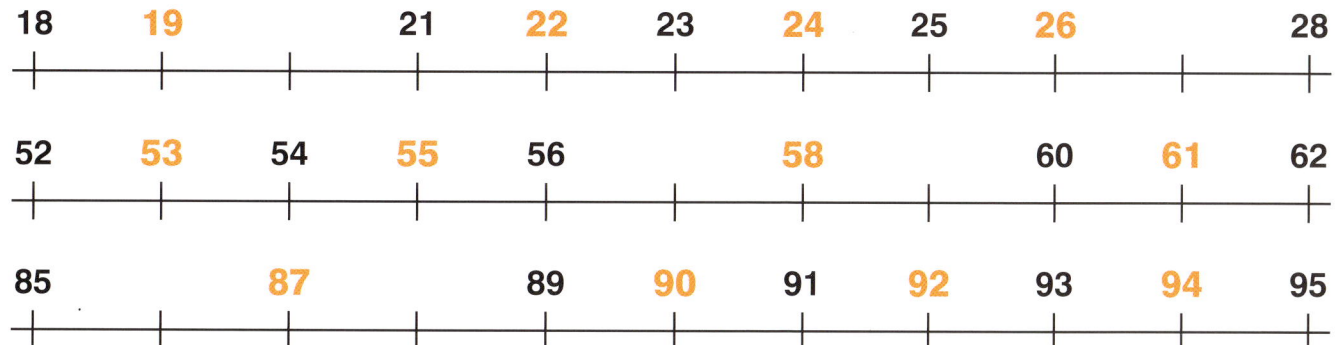

Paired activity 1A

Name:

My partner's name:

Date:

Compare and order two-digit numbers
and position them on a partially numbered
number line [lots of numbers already in place]

Speak Say each pair of numbers to your partner.

| 46 32 | 43 27 | 38 41 | 65 60 | 17 70 |

| 61 63 | 39 24 | 27 67 | 64 46 | 41 44 |

Listen Listen to your partner. They are going to say
10 different pairs of numbers.
Write the smaller number in each pair into a box below.

Look at the numbers you have written
in the 10 boxes above.
Write these numbers on the number lines.

37   40   42   47

62   66   69   72

Compare your completed number lines with your partner's.
Are they the same? If not, find out why.

Paired activity 1B

Name:

My partner's name:

Date:

Compare and order two-digit numbers
and position them on a partially numbered
number line [lots of numbers already in place]

Listen Listen to your partner. They are going to say
10 different pairs of numbers.
Write the larger number in each pair into a box below.

Speak Say each pair of numbers to your partner.

39 55	72 65	44 84	72 46	80 63
75 70	43 53	76 67	64 94	41 43

Look at the numbers you have written
in the 10 boxes at the top of the sheet.
Write these numbers on the number lines.

```
37            40        42                          47
+---+---+---+---+---+---+---+---+---+---+
```

```
62                  66            69        72
+---+---+---+---+---+---+---+---+---+---+
```

Compare your completed number lines with your partner's.
Are they the same? If not, find out why.

Individual activity 2A

Compare and order two-digit numbers and position them on a partially numbered number line [fewer numbers already in place]

Write these numbers in order, smallest to largest.

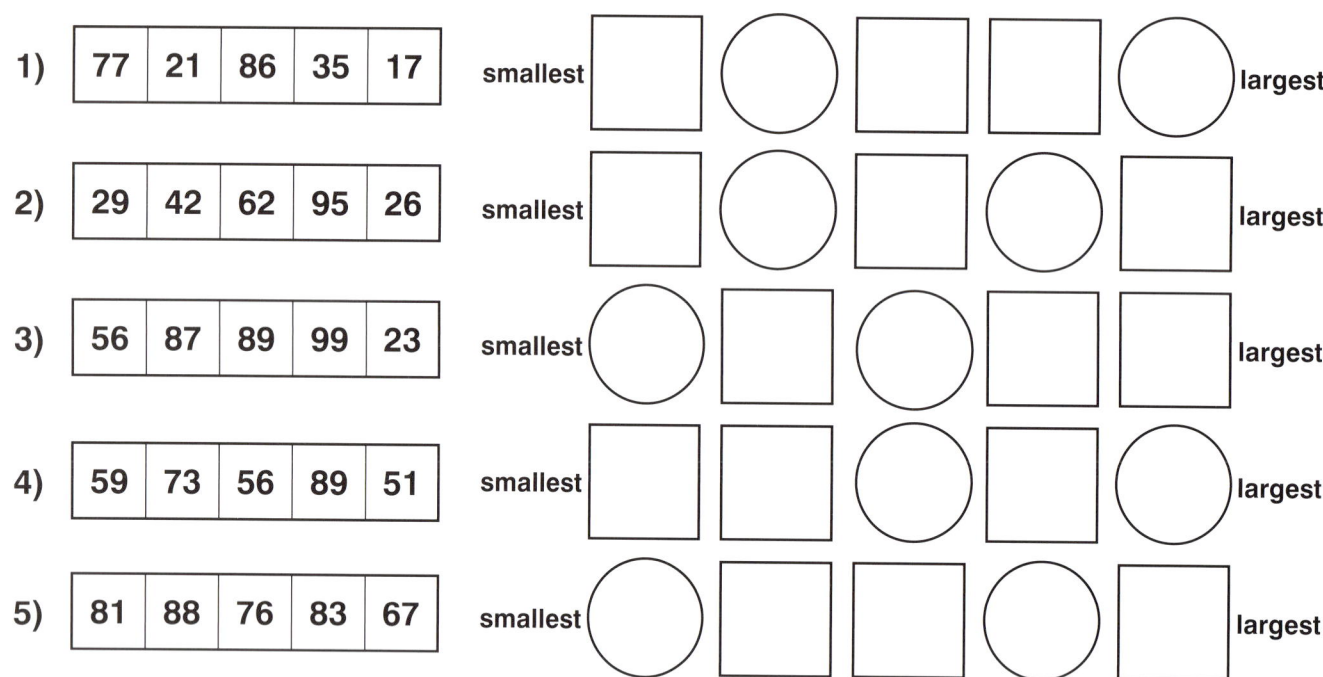

1) | 77 | 21 | 86 | 35 | 17 | smallest ... largest

2) | 29 | 42 | 62 | 95 | 26 | smallest ... largest

3) | 56 | 87 | 89 | 99 | 23 | smallest ... largest

4) | 59 | 73 | 56 | 89 | 51 | smallest ... largest

5) | 81 | 88 | 76 | 83 | 67 | smallest ... largest

6) | 76 | 17 | 66 | 27 | 65 | smallest ... largest

Look at the 12 numbers you have written in the circles ◯.
Write these numbers on the number lines.

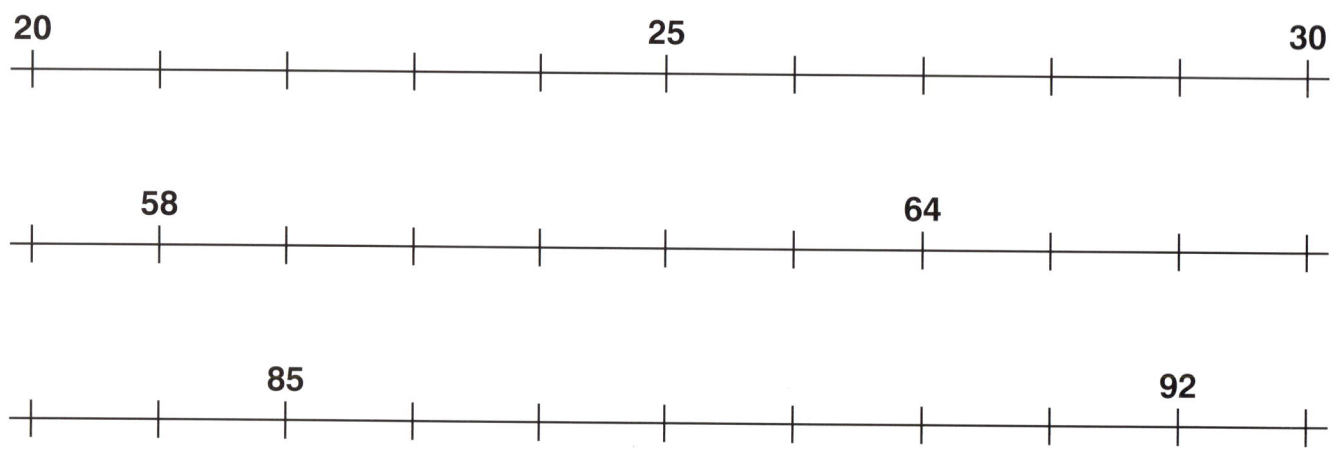

20 .... 25 30

58 .... 64 ....

85 .... 92 ....

Now compare your completed number lines with your partner's.
Are they the same? If not, find out why.

Individual activity 2B

Name: _____

Date: _____

Compare and order two-digit numbers and position them on a partially numbered number line [fewer numbers already in place]

Write these numbers in order, smallest to largest.

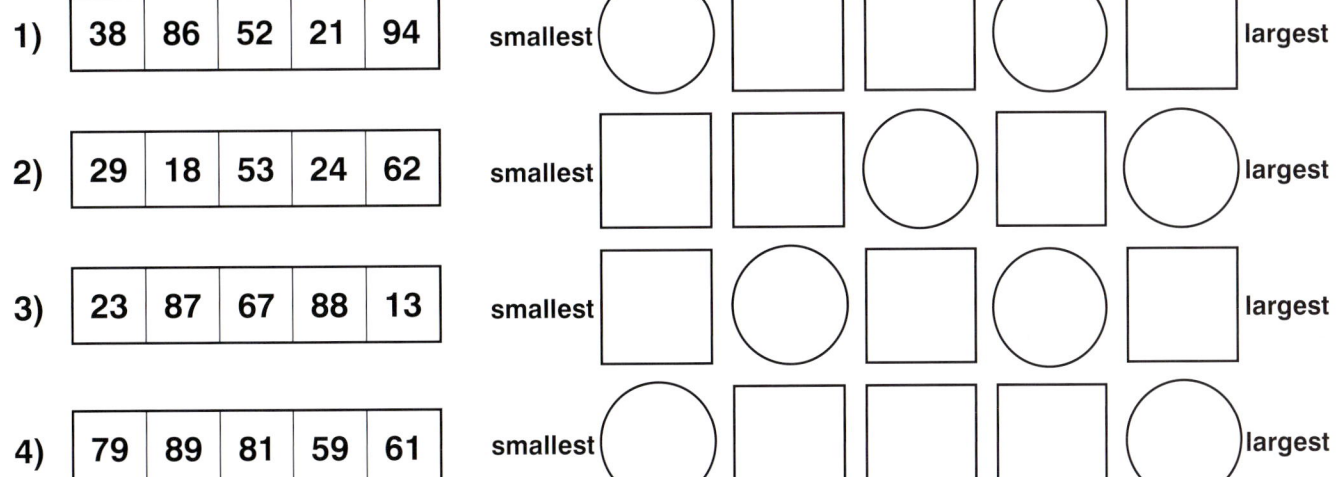

1) | 38 | 86 | 52 | 21 | 94 | smallest ◯ □ □ ◯ □ largest

2) | 29 | 18 | 53 | 24 | 62 | smallest □ □ ◯ □ ◯ largest

3) | 23 | 87 | 67 | 88 | 13 | smallest □ ◯ □ ◯ □ largest

4) | 79 | 89 | 81 | 59 | 61 | smallest ◯ □ □ □ ◯ largest

5) | 67 | 42 | 98 | 88 | 83 | smallest □ ◯ ◯ □ □ largest

6) | 40 | 72 | 23 | 65 | 27 | smallest □ ◯ □ ◯ □ largest

Look at the 12 numbers you have written in the circles ◯.
Write these numbers on the number lines.

20   25   30

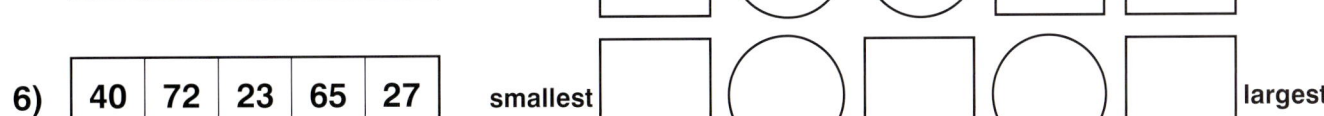

58   64

85   92

Now compare your completed number lines with your partner's.
Are they the same? If not, find out why.

Paired activity 2A

Name: _____

My partner's name: _____

Date: _____

Compare and order two-digit numbers and position them on a partially numbered number line [fewer numbers already in place]

You need:
coloured pencil

Speak Say the box number and the three numbers
in each of the triangles △ to your partner.

Box 1

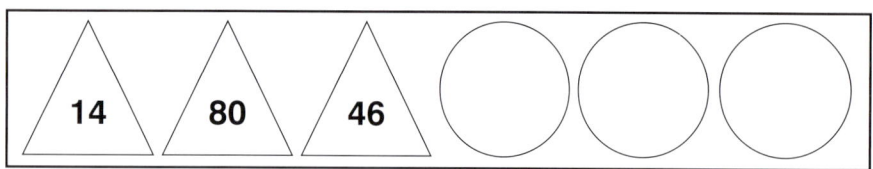

Box 2

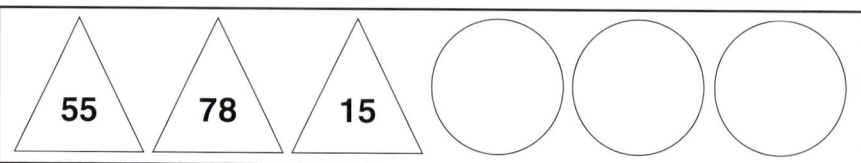

Box 3

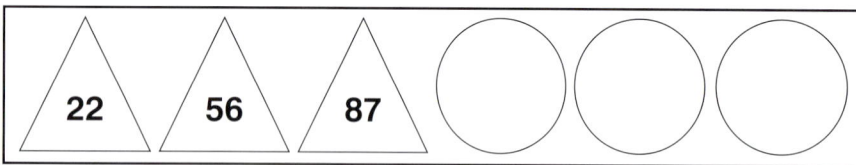

Box 4

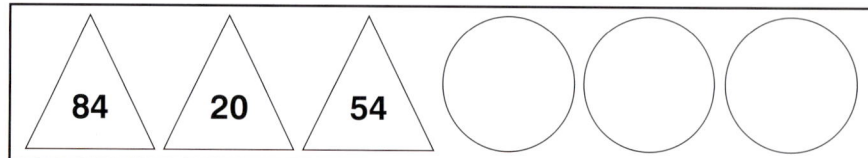

Listen Listen to you partner and write the three numbers
they tell you in the circles ◯.
Make sure you write the numbers in the correct box.

Now colour the smallest and largest number in each box.

Compare the numbers you have coloured
with the numbers your partner has coloured.
They should be the same. If not, find out why.

Work with your partner to write the eight numbers
you have coloured on these number lines.

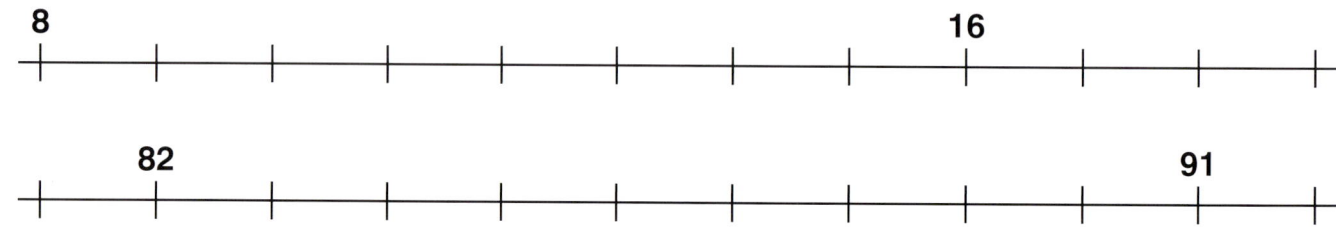

Name:

My partner's name:

Date:

Paired activity 2B

Compare and order two-digit numbers and position them on a partially numbered number line [fewer numbers already in place]

You need:
coloured pencil

Listen Listen to you partner and write the three numbers they tell you in the circles ◯.
Make sure you write the numbers in the correct box.

Box 1

◯ ◯ ◯ △52 △16 △86

Box 2

◯ ◯ ◯ △49 △19 △89

Box 3

◯ ◯ ◯ △12 △79 △48

Box 4

◯ ◯ ◯ △83 △51 △17

Speak Say the box number and the three numbers in each of the triangles △ to your partner.

Now colour the smallest and largest number in each box.

Compare the numbers you have coloured with the numbers your partner has coloured. They should be the same. If not, find out why.

Work with your partner to write the eight numbers you have coloured on these number lines.

8 ———————————————— 16 ————————

——— 82 ———————————————— 91 ———

Section 2

Place value and partitioning

Level 1
- Partition two-digit numbers into multiples of 10 and 1

Level 2
- Partition two-digit numbers in different ways, including into multiples of 10 and 1

Strategic approaches to develop fluency in place value and partitioning

 **Have a place value chart on the wall and frequently rehearse fluent naming of numbers**

The place value chart was invented by Caleb Gattegno, a mathematics educator, and is sometimes referred to as a Gattegno chart. It looks like this:

10	20	30	40	50	60	70	80	90
1	2	3	4	5	6	7	8	9

Just as numbers increase in size by a factor of 10 as you move up a row on the Gattegno chart, so numbers decrease by a factor of 10 as you move down a row.

Tapping out a number from the top row and then the bottom row, children can either individually or as a group say the number. You can also say a number and invite children to come tap out the number.

You could place this on the wall near the door so that children have to say or tap a number individually as they leave the class. Do such activities often for short bursts of time, rather than think of them as something you might do in only a few lessons for longer.

 Link the place value chart with base-10 blocks

Have some base-10 blocks on display below the place value chart. Circle two numbers on the chart, one in each row. Over the course of a day, children use the base-10 blocks in different ways to represent the number shown by the circles. Once they have had their model checked by a partner, they put the blocks back for the next person.

10	20	30	40	(50)	60	70	80	90
1	(2)	3	4	5	6	7	8	9

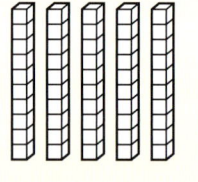

$52 = 50 + 2$

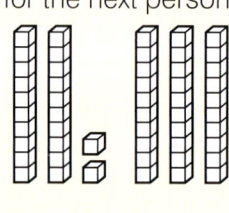

$52 = 22 + 30$

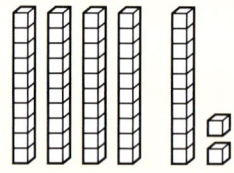

$52 = 40 + 12$

 Place value cards

Place value cards are another useful visual image in demonstrating to children how to partition numbers.

$$50 + 2 = 52$$

Place value grid

It is important that children make the connection between each of the digits in a two-digit number and their place value. Using a place value grid is a powerful visual image in demonstrating this.

The place value grid, like the Gattegno chart, also shows how multiplying by 10 moves a digit one place to the left and dividing by 10 moves a digit one place to the right.

Tens	Units

Expanded notation

When working with a Gattegno chart, place value cards or a place value grid to demonstrate place value and partitioning, also show the children how they can write numbers using expanded notation: for example, 52 can be written as $50 + 2$.

Once children are secure with partitioning two-digit numbers into tens and units, show how they can partition numbers in different ways.

$$40 + 12 \qquad 30 + 22 \qquad 20 + 32 \qquad 10 + 42$$

Individual and paired activities

Level 1 Partition two-digit numbers into multiples of 10 and 1

The individual activity provides practice in writing partitioned two-digit numbers using the Gattegno chart.

What number is shown on this chart?

Point to the number 48 on this chart.

The paired activity provides practice in partitioning two-digit numbers into multiples of 10 and 1, using place value cards as a visual image. Children draw lines to match multiples of 10 and 1 to make different two-digit numbers. If both children are correct in their partitioning, the lines should form the same pattern on both sheets.

Which multiple of 10 and another number make up the number 63?

What two-digit number do 80 and 3 make?

Level 2 Partition two-digit numbers in different ways, including into multiples of 10 and 1

In the individual activity, children complete pairs of calculations that show partitioned two-digit numbers, using expanded notation. When they have completed the activity, they compare answers. If they are both correct, the missing numbers in each calculation should be the same on both sheets.

What is the missing number from this number sentence? How do you know?

Can you see any pattern in each pair of number sentences?

The paired activity provides practice in partitioning two-digit numbers into multiples of 10 and 1, using the Gattegno chart. When children have partitioned their numbers, they work with their partner to split each of their numbers in other ways.

54 is the same as 50 add 4. How else can you partition 54? Can you tell me another way?

How did you partition 95? How else could you do it?

Further activities to develop fluency

Larger number wins

Provide each child with a sheet similar to the one on the right and each pair with a set of 0–9 digit cards.

Children shuffle the cards and place them in a pile on the table, face down. They take turns to take the top card from the pack and position it in one of the columns on their grid or put it in the dustbin. Once they have placed a card in a position, they cannot move it, and one number only is allowed in the dustbin.

Children continue taking cards until they have made a two-digit number. Each child reads out the number they have made. The winner of the round is the player with the larger number.

They then collect the cards, reshuffle them and place them back into a pile. The overall winner is the player who wins the most rounds.

Variation: Children play 'Smaller number wins'

Tens	Units

Representing place value

Provide each pair or group with a set of place value cards, a place value grid, a set of 0–9 digit cards, a Gattegno chart, some counters, base-10 blocks and paper and pencil.

The children work together to represent a two-digit number, using each of the different forms of apparatus, and write it, using expanded notation: for example, 52.

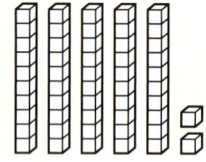

Tens	Units
5	**2**

10	20	30	40	(50)	60	70	80	90
1	(2)	3	4	5	6	7	8	9

$$52 = 50 + 2$$
$$= 30 + 22$$
$$= 20 + 32$$
$$= 10 + 42$$

Please give me your …

Provide each child with a calculator. Each child enters a two-digit number into their calculator. They then take turns to ask their partner for a particular place value part of the number that their partner currently has showing on the calculator: for example, Child A enters 35, and Child B enters 46. Child A says: "Please give me your tens." Child B says: "40", and Child A adds that to their number (so they now have 75). Then Child B says: "Please

give me your tens." Child A says: "30", and Child B adds that to their number (so they now have 76).

Children repeat for the other place value part of the number. They then compare calculator displays. Do both children end up with the same number (in the example above, 81)?

Children repeat several times.

My record sheet

Name: _____

Date: _____

	Before the activities			After the activities		
I can read numbers to 100 correctly.	☺	😐	☹	☺	😐	☹
I can explain what each digit in a two-digit number stands for.	☺	😐	☹	☺	😐	☹
I can split a number into tens and units: for example, $63 = 60 + 3$	☺	😐	☹	☺	😐	☹
I can split a number into tens and units in different ways: for example, $63 = 50 + 13$	☺	😐	☹	☺	😐	☹

After the activities

These numbers all have 4 tens.	
These numbers all have 3 units.	
I can split the number 46 in all these different ways.	
I can split the number 87 in all these different ways.	

Individual activity 1A

Partition two-digit numbers into multiples of 10 and 1

For each place value chart, combine the two numbers in the black boxes to make a two-digit number.
Write the number in the box ☐ on the left.

Then do the same for the two numbers in the grey boxes and write this number in the box ☐ on the right.
The first chart has been done for you.

1)

10	20	**30**	40	50	60	70	80	90
1	2	3	4	5	6	**7**	8	9

37   62

2)

10	**20**	30	40	50	60	70	80	90
1	2	3	4	5	6	7	**8**	9

☐   ☐

3)

10	20	30	40	50	60	70	80	90
1	2	**3**	4	5	6	7	8	9

☐   ☐

4)

10	20	30	40	**50**	60	70	80	90
1	2	3	4	5	6	7	8	9

☐   ☐

5)

10	20	30	40	50	60	70	**80**	90
1	2	3	4	5	6	7	**8**	9

☐   ☐

6)

10	20	30	40	50	**60**	70	80	90
1	2	**3**	4	5	6	7	8	9

☐   ☐

Answers to 1B

1) ☐37☐   ☐62☐   2) ☐87☐   ☐53☐   3) ☐95☐   ☐12☐

4) ☐38☐   ☐46☐   5) ☐24☐   ☐61☐   6) ☐49☐   ☐77☐

Individual activity 1B

Name:

Date:

Partition two-digit numbers into multiples of 10 and 1

For each place value chart, combine the two numbers in the black boxes to make a two-digit number. Write the number in the box ☐ on the left.

Then do the same for the two numbers in the grey boxes and write this number in the box ☐ on the right. The first chart has been done for you.

1)
20	**30**	40	50	60	70	80	90	90
2	3	4	5	6	**7**	8	9	9

☐ 37   ☐ 62

2)
10	20	30	40	50	60	70	**80**	90
1	2	3	4	5	6	**7**	8	9

☐   ☐

3)
10	20	30	40	50	60	70	80	**90**
1	2	3	4	**5**	6	7	8	9

☐   ☐

4)
10	20	**30**	40	50	60	70	80	90
1	2	3	4	5	6	7	**8**	9

☐   ☐

5)
10	**20**	30	40	50	60	70	80	90
1	2	3	**4**	5	6	7	8	9

☐   ☐

6)
10	20	30	**40**	50	60	70	80	90
1	2	3	4	5	6	7	8	**9**

☐   ☐

Answers to 1A

1) ☐ 37   ☐ 62   2) ☐ 28   ☐ 45   3) ☐ 13   ☐ 76

4) ☐ 59   ☐ 94   5) ☐ 88   ☐ 35   6) ☐ 63   ☐ 21

Paired activity 1A

Name:

My partner's name:

Date:

Partition two-digit numbers into multiples of 10 and 1

Speak Slowly say these numbers to your partner.

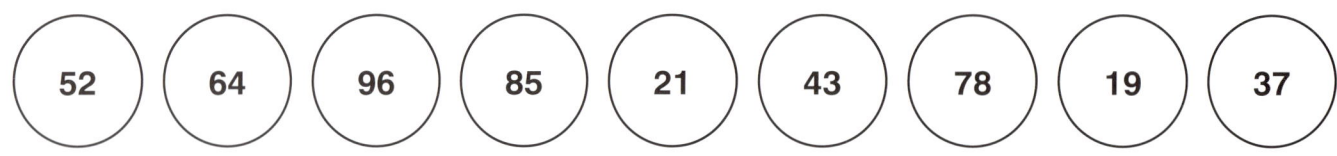

52 64 96 85 21 43 78 19 37

Listen Listen to the numbers your partner says.
Split each number into tens and units and draw a line to join together the tens and the units.

10	2
60	5
40	9
20	3
90	6
30	8
70	4
50	7
80	1

Look at the numbers in the circles ◯ on your partner's sheet to check if you correctly joined together the tens and the units.

Compare the pattern you made from joining the tens to the units to the pattern your partner made. What do you notice?

Paired activity 1B

Name:

My partner's name:

Date:

Partition two-digit numbers into multiples of 10 and 1

Listen Listen to the numbers your partner says.
Split each number into tens and units and draw a line to join together the tens and the units.

2 0	4
1 0	2
6 0	5
8 0	1
4 0	8
5 0	9
9 0	3
7 0	7
3 0	6

Speak Slowly say these numbers to your partner.

(42) (71) (13) (94) (35) (68) (56) (29) (87)

Look at the numbers in the circles ◯ on your partner's sheet to check if you correctly joined together the tens and the units.

Compare the pattern you made from joining the tens to the units to the pattern your partner made. What do you notice?

Individual activity 2A

Partition two-digit numbers in different ways, including into multiples of 10 and 1

Complete each pair of number sentences.

$36 = 30 + \boxed{}$

$36 = 20 + \boxed{}$

$78 = 70 + \boxed{}$

$78 = 50 + \boxed{}$

$43 = \boxed{} + 3$

$43 = \boxed{} + 13$

$74 = \boxed{} + 24$

$74 = \boxed{} + 44$

$82 = 70 + \boxed{}$

$82 = \boxed{} + 42$

$29 = \boxed{} + 19$

$29 = 20 + \boxed{}$

Compare the numbers in the boxes with your partner's.
Are they the same? If not, find out why.

Name: _____

Date: _____

Partition two-digit numbers in different ways, including into multiples of 10 and 1

Complete each pair of number sentences.

56 = 50 + [ ]

56 = 40 + [ ]

48 = 40 + [ ]

48 = 20 + [ ]

83 = [ ] + 43

83 = [ ] + 53

54 = [ ] + 4

54 = [ ] + 24

92 = 80 + [ ]

92 = [ ] + 52

69 = [ ] + 59

69 = 60 + [ ]

Compare the numbers in the boxes with your partner's.
Are they the same? If not, find out why.

Paired activity 2A

Name: _____

My partner's name: _____

Date: _____

Partition two-digit numbers in different ways, including into multiples of 10 and 1

You need:
coloured pencil

Speak Slowly say each of the numbers in the circles ◯ to your partner.

Do not say the number sentences in the grey boxes ▢.

1)
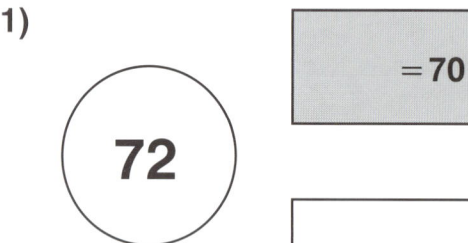

$= 70 + 2$

72

2)

$= 50 + 4$

54

3)
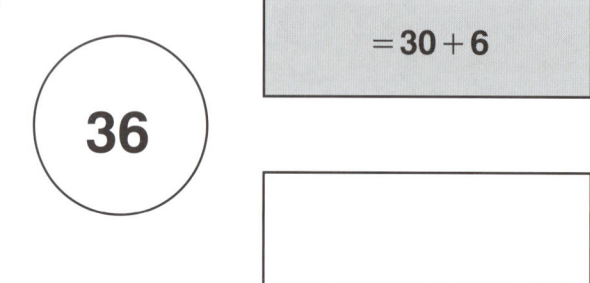

$= 30 + 6$

36

4)

$= 80 + 7$

87

Listen Listen to the numbers your partner tells you.
Split each number into tens and units
and colour them on the place value charts.

1)
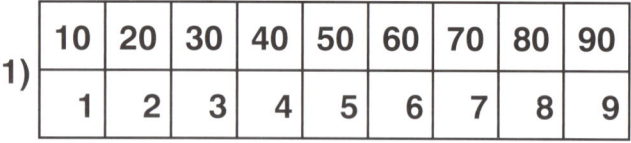

10	20	30	40	50	60	70	80	90
1	2	3	4	5	6	7	8	9

2)

10	20	30	40	50	60	70	80	90
1	2	3	4	5	6	7	8	9

3)
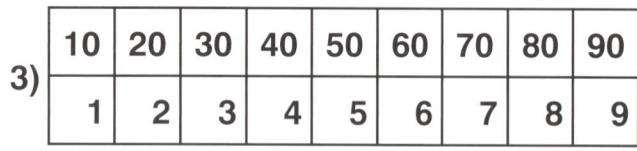

10	20	30	40	50	60	70	80	90
1	2	3	4	5	6	7	8	9

4)

10	20	30	40	50	60	70	80	90
1	2	3	4	5	6	7	8	9

Check your answers with your partner.
Your answers are in the grey boxes at the bottom
of your partner's sheet.

Now work with your partner to split each of your numbers in another way.

Write these in the white ▢ boxes above.

Paired activity 2B

**Partition two-digit numbers
in different ways, including into multiples of 10 and 1**

You need:
coloured pencil

Listen Listen to the numbers your partner tells you.
Split each number into tens and units
and colour them on the place value charts.

1)

10	20	30	40	50	60	70	80	90
1	2	3	4	5	6	7	8	9

2)

10	20	30	40	50	60	70	80	90
1	2	3	4	5	6	7	8	9

3)

10	20	30	40	50	60	70	80	90
1	2	3	4	5	6	7	8	9

4)

10	20	30	40	50	60	70	80	90
1	2	3	4	5	6	7	8	9

Speak Slowly say each of the numbers in the circles ◯ to your partner.
Do not say the number sentences in the grey boxes ▭.

1)

43 $= 40 + 3$

2)

69 $= 60 + 9$

3)

95 $= 90 + 5$

4)

38 $= 30 + 8$

Check your answers with your partner.
Your answers are in the grey boxes at the top
of your partner's sheet.

Now work with your partner to split each of your numbers in another way.
Write these in the white boxes ▭ above.

Understanding and using tens

Level 1
- Add mentally a one-digit number to any two-digit number

Level 2
- Add mentally a one-digit number to any two-digit number and a multiple of 10 to any two-digit number

Strategic approaches to develop fluency in adding mentally calculations such as TU + U and TU + T

 ## Use known facts to derive unknown facts

Show children the link between calculations such as 35 + 60 and 30 + 60 (using knowledge of 3 + 6) and calculations such as 47 + 8 and 7 + 8. Encourage the children to use known facts to help derive answers to unknown facts.

When adding a multiple of 10 to a two-digit number (such as 35 + 60) and a one-digit number to a two-digit number (such as 47 + 8), this involves being able to use and apply knowledge of place value and partitioning (for example, 35 = 30 + 5), recall addition number facts, and add together pairs of multiples of 10 such as 30 and 60.

 ## Focus on the commutative rule and putting the larger number first

Encourage children not simply to carry out calculations in the order in which they are presented on the page: for example, given

35 + 60, it is much easier to calculate 60 + 35. Similarly, 47 + 8 is easier to calculate than 8 + 47.

 ## Think, choose and use

Encourage children to think about different ways of answering a calculation before choosing one to use. Ask questions such as: "How could you work this out? How else might you work it out? Is there another way?" Discuss various suggestions offered by the children.

There is no 'proper' method of doing a calculation. What is important is that you teach children to use the most effective and efficient method for them: in other words, a method that always works (effective) and works quickly (efficient).

 ## Model strategies on an empty number line

An empty number line provides a visual model of different mental strategies:

35 + 60

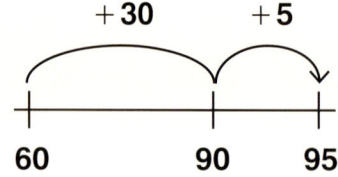

47 + 8

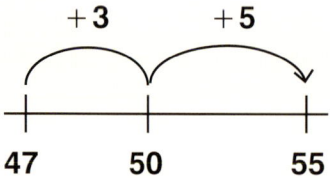

Individual and paired activities

 Level 1 Add mentally a one-digit number to any two-digit number

In the individual activity, children add one-digit numbers to two-digit numbers. They then look at their answers and draw lines between calculations with the same answer. If both children are correct, they should have the same answers when they compare sheets and see the same pattern of lines.

What is 78 add 6? How did you work that out?

Could you have worked it out using a different method?

In the paired activity, children draw lines between different pairs of numbers, a one-digit number and a two-digit number, then add the two numbers together. Working with their partner, they each draw lines between different pairs of numbers and work out the answers. When they have finished answering the questions, both children should have the same answers if they are correct.

Did you and your partner end up with the same answer? Why not?

What is the sum of 68 and 9? Did you both work it out the same way?

 Level 2 Add mentally a one-digit number to any two-digit number and a multiple of 10 to any two-digit number

In the individual activity, children answer a set of calculations and use their answers to complete a 1–100 grid dot-to-dot puzzle. Both children's picture should be the same.

How did you work out the answer to 49 add 9?

In the paired activity, children take turns to say a number. Both children use the two numbers to make a calculation. Children then work out the answers to the calculations. If they are both correct, the children should have the same answers.

Did you both get the same answer to 40 add 59? Did you both work it out the same way?

1	2	3	4	5	6	7	8	9	10
11	12	13	14	15	16	17	18	19	20
21	22	23	24	25	26	27	28	29	30
31	32	33	34	35	36	37	38	39	40
41	42	43	44	45	46	47	48	49	50
51	52	53	54	55	56	57	58	59	60
61	62	63	64	65	66	67	68	69	70
71	72	73	74	75	76	77	78	79	80
81	82	83	84	85	86	87	88	89	90
91	92	93	94	95	96	97	98	99	100

Further activities to develop fluency

 **How many ways?**

Children write down as many different addition calculations with an answer of, for example, 23.

Making a true calculation

Write a set of calculations similar to the following on the board:

6 [ ] + 7 = [ ]

[ ] 3 + 9 = [ ]

[ ] 6 + [ ] = [ ]

7 [ ] + [ ] = [ ]

[ ] 8 + 4 0 = [ ]

4 [ ] + 3 0 = [ ]

[ ] 3 + [ ] 0 = [ ]

7 [ ] + [ ] 0 = [ ]

Can the children suggest digits for each box so that the calculations are correct?
What other digits could go in the boxes?

Target board

Display a target board containing a combination of one-digit and two-digit numbers and multiples of 10.

Ask questions to the class that require children to add pairs of numbers on the board.

What is the sum of 28 and 6?

What is 60 add 14?

What two numbers next to each other on the board make a total of 84?

Tell me an addition number sentence involving two of the numbers on the board.

19	70	9	73	60
6	82	61	30	25
28	5	37	7	55
20	94	80	36	14
76	8	46	69	4
50	47	58	90	93

The adding game

Give each pair a pack of 1–9 digit cards. Children shuffle the cards and place them in a pile, face down. Children take three cards each and arrange them to make a two-digit and a one-digit number: for example, 54 9. They add their two numbers together.

The child with the highest total is the winner.

Collect the cards, reshuffle them and place them back in a pile. Play 8 rounds.

Variations:

The winner is the player with:

• the lowest total

• the lowest even total

• the highest odd total

• the nearest total to 60.

My record sheet

Name:

Date:

	Before the activities	After the activities
I can add a one-digit number to a two-digit number in my head: for example, 26 + 8.	☺ 😐 ☹	☺ 😐 ☹
I can add a two-digit number to a multiple of 10 in my head: for example, 23 + 60.	☺ 😐 ☹	☺ 😐 ☹

◉ After the activities

Here are some addition calculations such as 26 + 8 that I can answer in my head.

☐☐ + ☐ = ☐☐   ☐☐ + ☐ = ☐☐☐

☐☐ + ☐ = ☐☐   ☐☐ + ☐ = ☐☐☐

Here are some addition calculations such as 23 + 60 that I can answer in my head.

☐☐ + ☐☐ = ☐☐

☐☐ + ☐☐ = ☐☐

☐☐ + ☐☐ = ☐☐

☐☐ + ☐☐ = ☐☐

Individual activity 1A

Name:

Date:

Add mentally a one-digit number to any two-digit number

Work out the answer to each of the number sentences.

23 + 8 = ⬜ •

42 + 3 = ⬜ •

7 + 34 = ⬜ •

4 + 56 = ⬜ •

47 + 5 = ⬜ •

67 + 6 = ⬜ •

8 + 76 = ⬜ •

• 5 + 36 = ⬜

• 51 + 9 = ⬜

• 39 + 6 = ⬜

• 8 + 65 = ⬜

• 6 + 78 = ⬜

• 27 + 4 = ⬜

• 7 + 45 = ⬜

Look at the numbers you have written in the boxes ⬜.
Draw lines between dots that have the same answer.

Compare your sheet with your partner's sheet.
Do you have the same answers and the same pattern of lines?
If not, find out why.

Add mentally a one-digit number to any two-digit number

Work out the answer to each of the number sentences.

24 + 7 =

41 + 4 =

8 + 33 =

5 + 55 =

46 + 6 =

64 + 9 =

7 + 77 =

6 + 35 =

53 + 7 =

37 + 8 =

5 + 68 =

9 + 75 =

26 + 5 =

4 + 48 =

Look at the numbers you have written in the boxes ☐.
Draw lines between dots that have the same answer.

Compare your sheet with your partner's sheet.
Do you have the same answers and the same pattern of lines?
If not, find out why.

Paired activity 1A

Add mentally a one-digit number to any two-digit number

Name: _____

My partner's name: _____

Date: _____

Draw lines to match a white number box ☐ to a black number box ■.

Add these two numbers together and write the total on the line joining the two numbers.
One has been done for you.

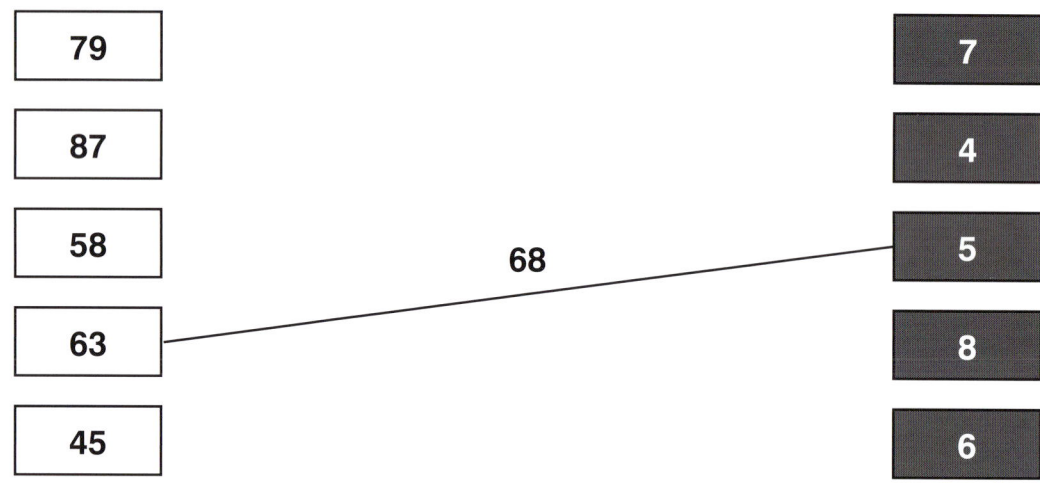

Speak Say each of the five pairs of numbers that are joined together to your partner, but not the total. For example, 63 and 5.

Listen Listen to your partner and draw lines to match each pair of numbers they tell you.

Now add together each of the above pairs of numbers and write the total on the line joining the two numbers.

Compare your answers with your partner.
Are they the same? If not, find out why.

Paired activity 1B

Name: _____

My partner's name: _____

Date: _____

**Add mentally a one-digit number
to any two-digit number**

Draw lines to match a white number box ☐ to a black number box ■.

Add these two numbers together and write the total
on the line joining the two numbers.
One has been done for you.

White		Black
35		6
68	74	5
47		4
84		9
56		8

Listen Listen to your partner and draw lines to match
each pair of numbers they tell you.

White	Black
79	7
87	4
58	5
63	8
45	6

Speak Say each of the five pairs of numbers that are joined together at the top
of the sheet to your partner, but not the total. For example, 68 and 6.

Now add together each of the above pairs of numbers and
write the total on the line joining the two numbers.

Compare your answers with your partner.
Are they the same? If not, find out why.

Individual activity 2A

Add mentally a one-digit number to any two-digit number and a multiple of 10 to any two-digit number

Answer each of these number sentences.

1) $1 + 2 = $ ☐

2) $14 + 9 = $ ☐

3) $16 + 5 = $ ☐

4) $30 + 23 = $ ☐

5) $69 + 4 = $ ☐

6) $24 + 40 = $ ☐

7) $30 + 46 = $ ☐

8) $77 + 20 = $ ☐

9) $72 + 6 = $ ☐

10) $71 + 9 = $ ☐

11) $28 + 30 = $ ☐

12) $30 + 8 = $ ☐

13) $20 + 10 = $ ☐

14) $11 + 5 = $ ☐

15) $2 + 3 = $ ☐

16) $2 + 1 = $ ☐

Now look at this number square.

Draw a line from the number that is the answer to Question 1, which is 3, to the number that is the answer to Question 2.

Continue the line to the number that is the answer to Question 3.

Keep going like this until you have joined all 16 answers.

What have you drawn?
Is it the same as your partner's?

1	2	3	4	5	6	7	8	9	10
11	12	13	14	15	16	17	18	19	20
21	22	23	24	25	26	27	28	29	30
31	32	33	34	35	36	37	38	39	40
41	42	43	44	45	46	47	48	49	50
51	52	53	54	55	56	57	58	59	60
61	62	63	64	65	66	67	68	69	70
71	72	73	74	75	76	77	78	79	80
81	82	83	84	85	86	87	88	89	90
91	92	93	94	95	96	97	98	99	100

Individual activity 2B

Name:

Date:

Add mentally a one-digit number to any two-digit number and a multiple of 10 to any two-digit number

Answer each of these number sentences.

1) $2 + 1 =$

2) $3 + 2 =$

3) $12 + 4 =$

4) $27 + 3 =$

5) $18 + 20 =$

6) $49 + 9 =$

7) $76 + 4 =$

8) $58 + 20 =$

9) $47 + 50 =$

10) $36 + 40 =$

11) $57 + 7 =$

12) $13 + 60 =$

13) $48 + 5 =$

14) $17 + 4 =$

15) $16 + 7 =$

16) $1 + 2 =$

Now look at this number square.

Draw a line from the number that is the answer to Question 1, which is 3, to the number that is the answer to Question 2.

Continue the line to the number that is the answer to Question 3.

Keep going like this until you have joined all 16 answers.

What have you drawn?
Is it the same as your partner's?

1	2	3	4	5	6	7	8	9	10
11	12	13	14	15	16	17	18	19	20
21	22	23	24	25	26	27	28	29	30
31	32	33	34	35	36	37	38	39	40
41	42	43	44	45	46	47	48	49	50
51	52	53	54	55	56	57	58	59	60
61	62	63	64	65	66	67	68	69	70
71	72	73	74	75	76	77	78	79	80
81	82	83	84	85	86	87	88	89	90
91	92	93	94	95	96	97	98	99	100

Paired activity 2A

Add mentally a one-digit number to any two-digit number and a multiple of 10 to any two-digit number

Speak Say the question number and the number in the square ☐ to your partner, like this. *Question 1, Number 7!*

Listen Your partner will say a number back to you.

Write this number in the circle ◯ beside the same question.

Do this until you have written a number in each circle ◯.

1) ☐ 7 + ◯ = △

2) ☐ 48 + ◯ = △

3) ☐ 30 + ◯ = △

4) ☐ 26 + ◯ = △

5) ☐ 64 + ◯ = △

6) ☐ 9 + ◯ = △

7) ☐ 43 + ◯ = △

8) ☐ 40 + ◯ = △

9) ☐ 27 + ◯ = △

10) ☐ 4 + ◯ = △

11) ☐ 60 + ◯ = △

12) ☐ 19 + ◯ = △

Now work out the answer to each number sentence and write it in the triangle △.

Compare your answers with your partner's and talk about any answers that are different.

Name:

My partner's name:

Date:

Paired activity 2B

Add mentally a one-digit number to any two-digit number and a multiple of 10 to any two-digit number

Listen Listen to your partner and write the number they say in the square □ beside the question they tell you.

Speak For the same question number, say the number in the circle ○ to your partner, like this. *Question 1, Number 32!*

Do this until you have written a number in each square □.

1) □ + (32) = △

2) □ + (5) = △

3) □ + (57) = △

4) □ + (50) = △

5) □ + (8) = △

6) □ + (82) = △

7) □ + (20) = △

8) □ + (59) = △

9) □ + (6) = △

10) □ + (38) = △

11) □ + (25) = △

12) □ + (70) = △

Now work out the answer to each number sentence and write it in the triangle △.

Compare your answers with your partner's and talk about any answers that are different.

Deriving and recalling addition and subtraction facts, and using that knowledge

Level 1
- Derive and recall all addition facts for each number to at least 10

Level 2
- Derive and recall all pairs of multiples of 10 with totals to 100

Strategic approaches to develop fluency in addition

Focus on the commutative rule for addition

There are 20 different addition number facts to learn for the numbers 1 to 5 and nearly 70 for the numbers 1 to 10.

Addition facts for 8
$8 + 0 = 8$
$7 + 1 = 8$
$6 + 2 = 8$
$5 + 3 = 8$
$4 + 4 = 8$
$3 + 5 = 8$
$2 + 6 = 8$
$1 + 7 = 8$
$0 + 8 = 8$

Becoming fluent in these is made a lot easier if children learn about the commutative property of addition sooner rather than later: that is, $3 + 5$ has the same answer as $5 + 3$.

Number tracks are a simple way of helping children understand this.

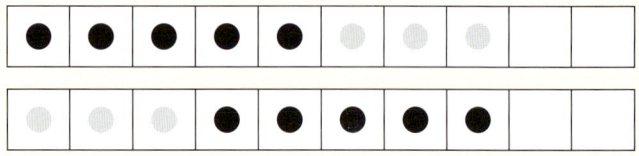

Using counters in two colours, put five in one colour on the top track, then three in the other colour. Put three in one colour on the second track and five in the other colour. Can the children explain why they get the same answer each time?

Put the larger number first

Children move though stages in becoming fluent in the number facts. First, they count all, so if they are figuring out $3 + 5$, they will count out three (using fingers or counters), count out another five and then count them all. Next, they start to count on: they count out five and then count on three from 5.

This second stage is easier if, building on the commutative law, you teach children to start with the larger number. You can easily assess if children are doing this by asking them what, for example, 7 plus 1 is, and what 1 plus 7 is. Children who answer each of these equally quickly have learnt that $7 + 1$ is a much easier calculation to do than $1 + 7$. Anyone who takes longer to answer the second question is probably counting on seven from 1. Recording the two different approaches on a number line or number track can help children appreciate the more effective approach.

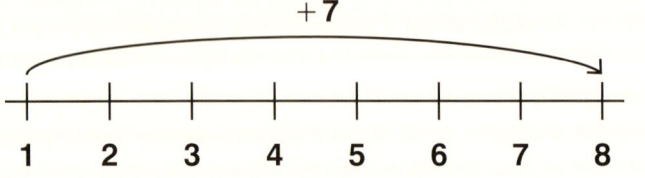

or

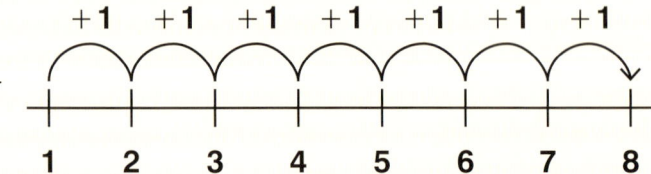

as opposed to the more efficient way:

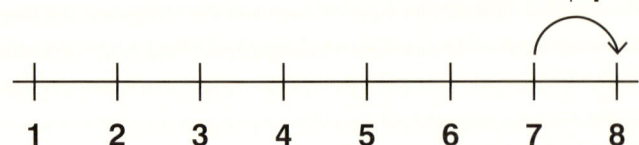

Inverse relationship and using 'number families' to work out subtraction facts

Some children find addition easier than subtraction and often use known addition facts to help them work out corresponding subtraction facts.

Building on the notion of the commutative rule for addition – if you know one addition fact, then you know another: for example, $5 + 3 = 8$ and $3 + 5 = 8$ – introduce the children to the inverse relationship between addition and subtraction and the notion of 'number families': if you know that $5 + 3 = 8$ and $3 + 5 = 8$, then you also know that $8 - 5 = 3$ and $8 - 3 = 5$. These four facts all contain the same three digits: 3, 5 and 8.

Using an array is a powerful model and image for demonstrating this.

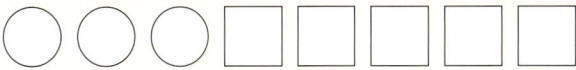

For the addition facts, ask:	For the subtraction, facts ask:
How many circles are there? (3) *How many squares are there?* (5) *How many shapes altogether?* (8) Write the calculation: $3 + 5 = 8$ *How many squares are there?* (5) *How many circles are there?* (3) *How many shapes altogether?* (8) Write the calculation: $5 + 3 = 8$	*How many shapes are there altogether?* (8) *If I were to take away the circles, how many shapes would I take away?* (3) *How many shapes would that leave me with?* (5) Write the calculation: $8 - 3 = 5$ *How many shapes are there altogether?* (8) *If I were to take away the squares, how many shapes would I take away?* (5) *How many shapes would that leave me with?* (3) Write the calculation: $8 - 5 = 3$

Individual and paired activities

Level 1 Derive and recall all addition facts for each number to at least 10

The individual activity provides practice in addition number facts to 10. Children work out the answers to a series of calculations, matching the first part of the number sentence with the answer. If both children are correct, they should have made the same patterns of lines when they compare sheets.

Tell me a number fact with an answer of 8. Can you tell me another one? Any more?

What is 3 add 7? Tell me another addition fact with a total of 10.

In the first part of the paired activity, children suggest pairs of numbers from 1 to 5 to add together. In the second part, they suggest pairs of numbers from 1 to 10 to add.

Tell me two numbers you added together. What is the answer?

Which set of six number sentences did you find the hardest? Why was this?

Level 2 Derive and recall all pairs of multiples of 10 with totals to 100

The individual activity provides practice in adding pairs of multiples of 10 with totals to 100. Children answer the 10 calculations, then join pairs of calculations with the same total.

What is 20 add 50? What other two multiples of 10 total 70?

In the paired activity, children work out the answer to six addition number facts to 10 and use them to derive and recall the totals of matching calculations involving multiples of 10. Children ask each other their 12 calculations to check.

What is 5 add 2? If 5 add 2 is 7, what is 50 add 20?

Further activities to develop fluency

 ## Minute calculations

One child rolls a 0–9 dice and sets a one-minute timer. All the children write down as many addition and/or subtraction calculations as possible that give the answer on the dice. After one minute, they compare and check each other's work. Each child's score for the round is the number of correct calculations they wrote down. The child who scores the highest rolls the dice for the next round.

The winner is the child with the highest score after five rounds.

 ## Trios

Prepare a set of demonstration cards that show an addition and subtraction 'number family'.

Hold up a card and hide one of the numbers.

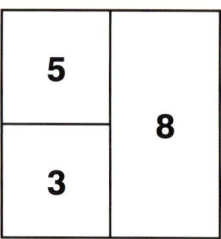
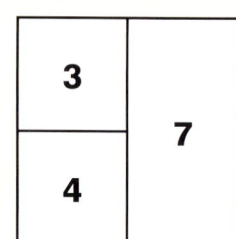

Children suggest different addition and subtraction calculations, using the two visible numbers.

What is 8 subtract 5?

5 and what other number makes 8?

What do you add to 5 to make 8?

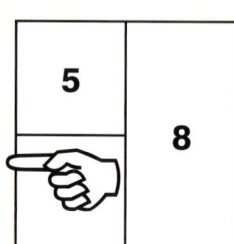

 ## Adding dice numbers

Provide each pair or group with two 1–6 or 0–9 dice and 20 counters. Children take turns to roll the dice. All the children add together the two numbers rolled. The first child to call out the correct answer takes a counter.

The winner is the child with more counters after 20 rounds.

 ## Adding multiples of 10

Provide each pair with a set of 1–9 digit cards. Children shuffle the cards and place them in a pile, face down. They turn over the top two cards such as 4 and 2 and make each number 10 times bigger: i.e. 40 and 20.

They then use this pair of numbers to make and answer an addition calculation: $40 + 20 = 60$

Children repeat this, making up 10 pairs of calculations.

 ## Multiples of 10 dominoes

Children choose a domino and use the numbers as single digits to make an addition calculation. They then make each number 10 times bigger and write the calculation: $40 + 20 = 60$

Children repeat this several times.

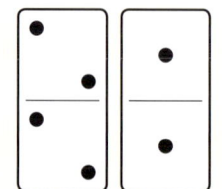

$4 + 2 = 6$

$40 + 20 = 60$

My record sheet

Name: _____

Date: _____

	Before the activities			After the activities		
I can add numbers to make different totals to 10.	☺	😐	☹	☺	😐	☹
I can add together pairs of multiples of 10 such as $40 + 30$.	☺	😐	☹	☺	😐	☹

After the activities

Here are some pairs of numbers that make 5.	
Here are some pairs of numbers that make 8.	
Here are some pairs of multiples of 10 added together.	

Individual activity 1A

Name:

Date:

Derive and recall all addition facts for each number to at least 10

Draw lines to match the addition with the right answer.
One has been done for you.

$3 + 4 =$	$2 + 3 =$	$5 + 1 =$	$7 + 3 =$	$4 + 5 =$	$4 + 4 =$
•	•	•	•	•	•

•	•	•	•	•	•
5	**6**	**7**	**8**	**9**	**10**
•	•	•	•	•	•

•	•	•	•	•	•
$3 + 5 =$	$3 + 3 =$	$1 + 4 =$	$3 + 6 =$	$5 + 2 =$	$6 + 4 =$

Compare your worksheet with your partner's.
If you are both correct, you should have
the same pattern of lines drawn.
If you don't, find out why.

Can you and your partner think of other additions
with answers of 5, 6, 7 …?
Write some of them on the back of this sheet.

Individual activity 1B

Name:

Date:

Derive and recall all addition facts for each number to at least 10

Draw lines to match the addition with the right answer.
One has been done for you.

$2+5=$	$4+1=$	$4+2=$	$5+5=$	$6+3=$	$5+3=$

5	6	7	8	9	10

$2+6=$	$3+3=$	$3+2=$	$7+2=$	$4+3=$	$8+2=$

Compare your worksheet with your partner's.
If you are both correct, you should have
the same pattern of lines drawn.
If you don't, find out why.

Can you and your partner think of other additions
with answers of 5, 6, 7 …?
Write some of them on the back of this sheet.

Paired activity 1A

Derive and recall all addition facts for each number to at least 10

Speak **Listen** Both you and your partner say a number from 1 to 5.
Write both these numbers in the circles ◯ in the first box.
Do this six times altogether.

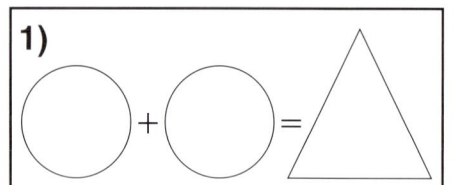

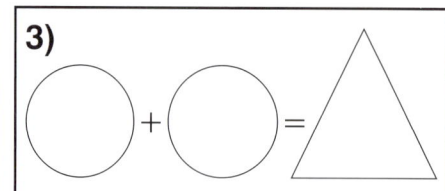

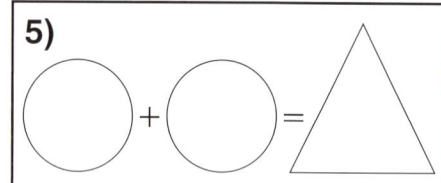

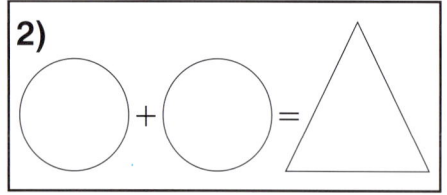

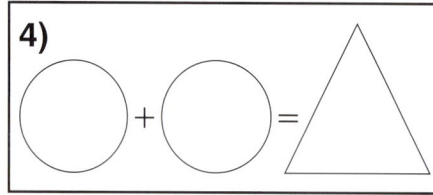

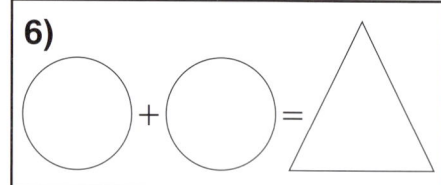

Speak **Listen** Say a number from 1 to 5.
This time, your partner is going to say a number from 6 to 10.
Write both these numbers in the circles ◯ in the first box.
Do this six times altogether.

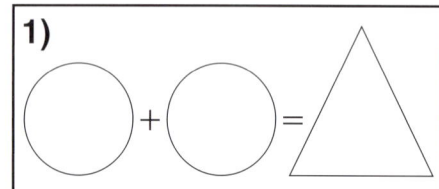

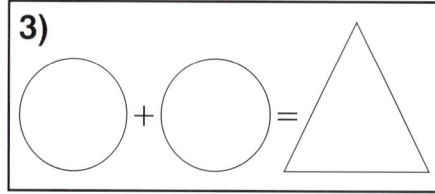

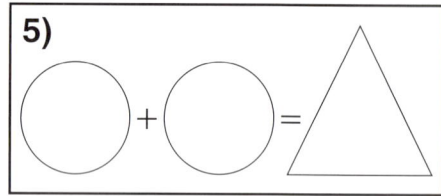

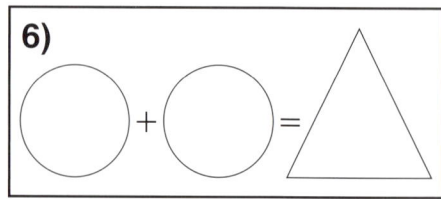

Complete each number sentence.

Compare your answers with your partner's.
Talk about any answers that are different.

Paired activity 1B

Name: _____

My partner's name: _____

Date: _____

Derive and recall all addition facts for each number to at least 10

Speak **Listen** Both you and your partner say a number from 1 to 5.

Write both these numbers in the circles ◯ in the first box.
Do this six times altogether.

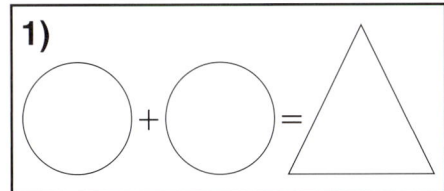

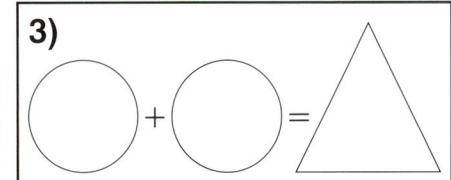

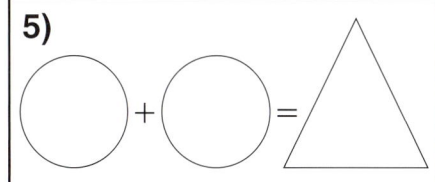

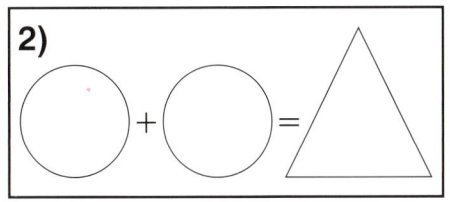

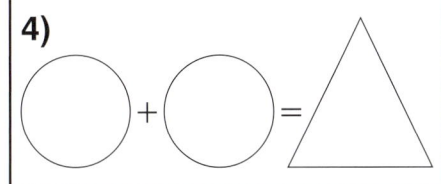

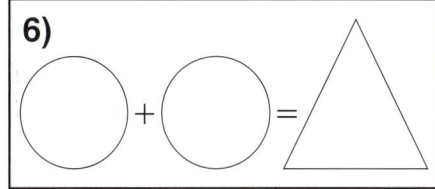

Speak **Listen** Your partner is going to say a number to you from 1 to 5.
This time, you say a number from 6 to 10.

Write both these numbers in the circles ◯ in the first box.
Do this six times altogether.

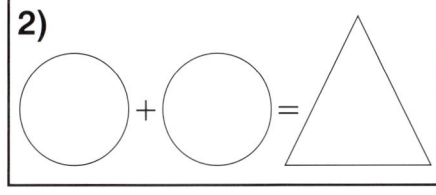

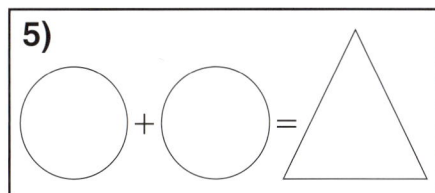

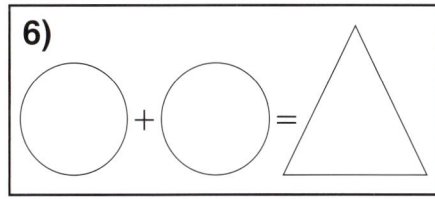

Complete each number sentence.

Compare your answers with your partner's.
Talk about any answers that are different.

Individual activity 2A

Name:

Date:

Derive and recall all pairs of multiples of 10 with totals up to 100

Work out the answer to each of the number sentences.
Two have been done for you.

1) 20 + 30 = **50**

2) 30 + 40 = ☐

3) 60 + 20 = ☐

4) 50 + 40 = ☐

5) 40 + 10 = **50**

6) 10 + 30 = ☐

7) 40 + 40 = ☐

8) 20 + 20 = ☐

9) 20 + 50 = ☐

10) 70 + 20 = ☐

Look at the numbers you have written in the answer boxes.
In each pair of circles, write the two question numbers
that show the same answer.
One has been done for you.

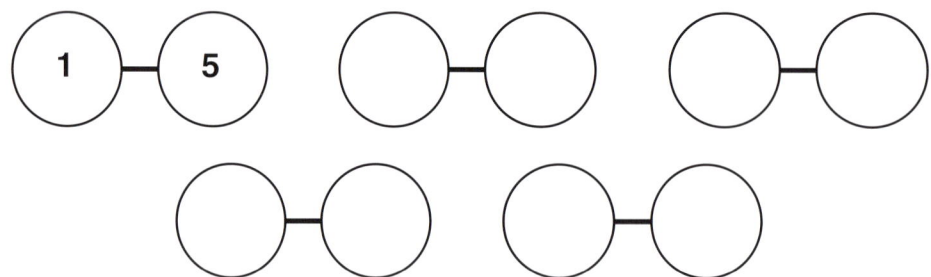

Check your answers with your partner.

Talk with your partner about how you
worked out some of your answers.

Answers to 1B

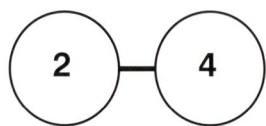 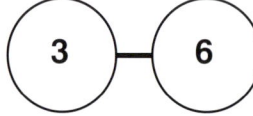 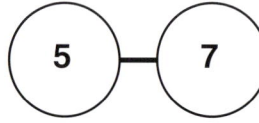 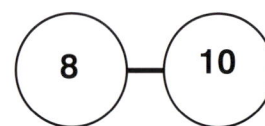

Individual activity 2B

Name: _____

Date: _____

Derive and recall all pairs of multiples of 10 with totals up to 100

Work out the answer to each of the number sentences.
Two have been done for you.

1) 30 + 30 = | 60 |

2) 50 + 30 = | |

3) 80 + 10 = | |

4) 20 + 60 = | |

5) 50 + 20 = | |

6) 60 + 30 = | |

7) 40 + 30 = | |

8) 30 + 20 = | |

9) 20 + 40 = | 60 |

10) 10 + 40 = | |

Look at the numbers you have written in the answer boxes.
In each pair of circles, write the two question numbers
that show the same answer.
One has been done for you.

(1)—(9)   ( )—( )   ( )—( )

( )—( )   ( )—( )

Check your answers with your partner.

Talk with your partner about how you
worked out some of your answers.

Answers to 1A

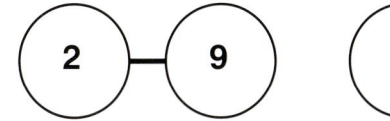 (2)—(9)   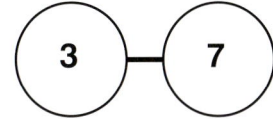 (3)—(7)   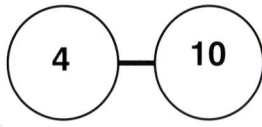 (4)—(10)    (6)—(8)

Paired activity 2A

Derive and recall all pairs of multiples of 10 with totals to 100

Write the answers in the white boxes.
Then draw lines to match each white box
to the related grey box.

Write the answers in the grey boxes.
One has been done for you.

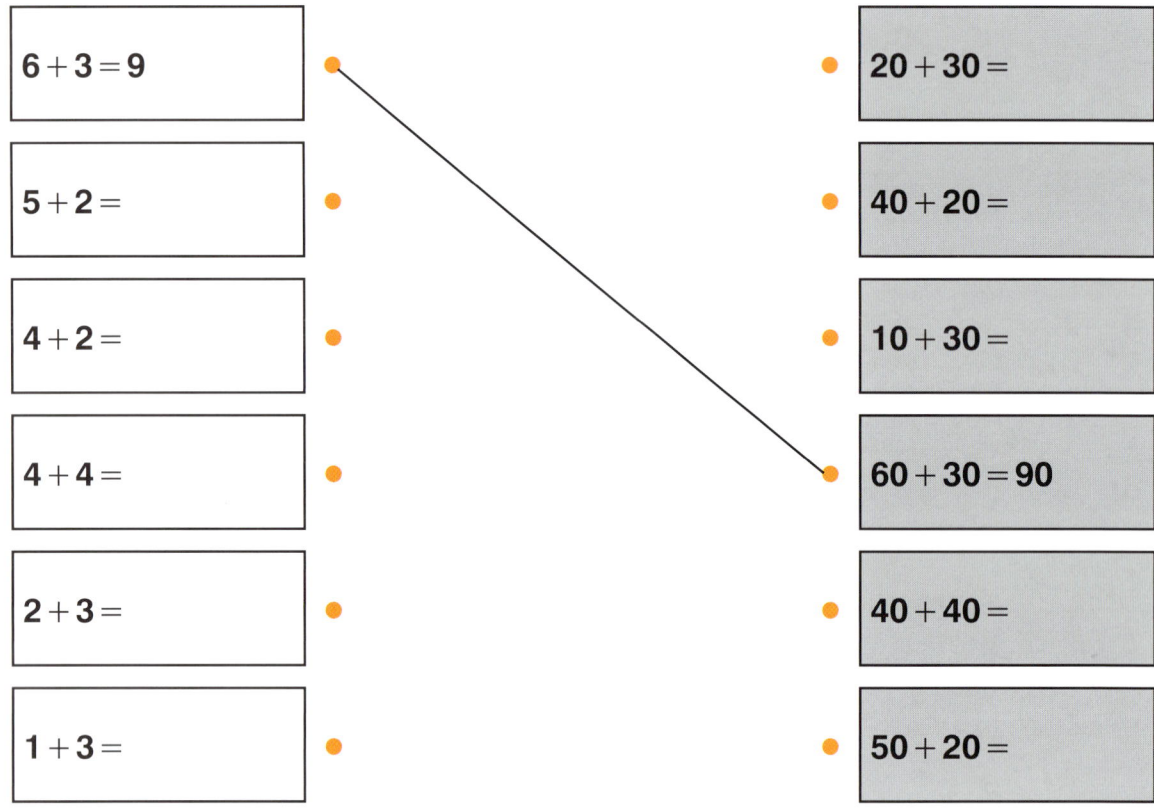

6 + 3 = 9		20 + 30 =
5 + 2 =		40 + 20 =
4 + 2 =		10 + 30 =
4 + 4 =		60 + 30 = 90
2 + 3 =		40 + 40 =
1 + 3 =		50 + 20 =

Speak Take turns to say one of the addition facts
in the white boxes and the related multiples of 10 fact
in the grey box to your partner. Don't say the answers.

Listen Listen to the two number sentences your partner
asks you and work out the answers.

Tell your partner the answers.
Do you both agree on the answers? If not, talk about it.

Continue until you have asked each other
all six pairs of number sentences.

Paired activity 2B

Derive and recall all pairs of multiples of 10 with totals to 100

Write the answers in the white boxes.
Then draw lines to match each white box
to the related grey box.

Write the answers in the grey boxes.
One has been done for you.

White boxes	Grey boxes
$2 + 7 = 9$	$30 + 20 =$
$4 + 3 =$	$10 + 50 =$
$1 + 5 =$	$20 + 20 =$
$3 + 5 =$	$20 + 70 = 90$
$3 + 2 =$	$30 + 50 =$
$2 + 2 =$	$40 + 30 =$

Speak Take turns to say one of the addition facts
in the white boxes and the related multiples of 10 fact
in the grey box to your partner. Don't say the answers.

Listen Listen to the two number sentences your partner
asks you and work out the answers.

Tell your partner the answers.
Do you both agree on the answers? If not, talk about it.

Continue until you have asked each other
all six pairs of number sentences.

Section 5

Deriving and recalling multiplication and division facts, and using that knowledge

Level 1
- Derive and recall multiplication facts for the 2, 5 and 10 times tables

Level 2
- Derive and recall division facts corresponding to the 2, 5 and 10 times tables

Strategic approaches to develop fluency in multiplication and division facts for the 2, 5 and 10 times tables

Focus on the commutative rule

There are 100 different multiplication facts to learn from 1×1 to 10×10.

Becoming fluent in these is made a lot easier if children learn about the commutative property of multiplication sooner rather than later: that is, 2×7 has the same answer as 7×2.

Arrays are a simple and powerful way of helping children understand this.

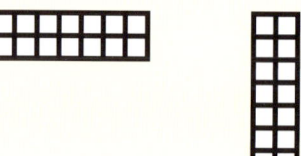

How many small squares in this array?

Without counting them all, how do we know this is also 14?

Use the 'key facts' of $1 \times$, $2 \times$, $5 \times$ and $10 \times$

Apart from being the easiest to learn and remember, there is another reason why children learn the 2, 5 and 10 times tables first. Once secure, they can use these facts to help them work out related facts.

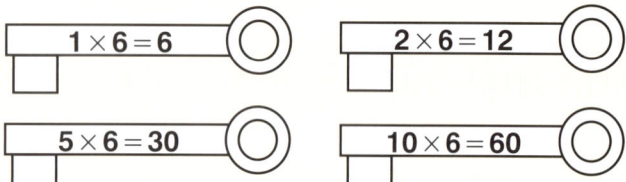

Having instant recall of these four facts can help derive answers to the following:

$3 \times 6 = (2 \times 6) + 6$ $\qquad$ $4 \times 6 = 2 \times 6$ **doubled**

$6 \times 6 = (5 \times 6) + 6$ $\qquad$ $7 \times 6 = (5 \times 6) + (2 \times 6)$

$8 \times 6 = (10 \times 6) - (2 \times 6)$ $\qquad$ $9 \times 6 = (10 \times 6) - 6$

Use doubling (the 2 times tables)

Children quickly learn the doubles of 1 to 5 (it's in the fingers) and then doubling all numbers to 10. This can help reduce the memory load on becoming fluent in multiplication. It is for this reason that you need to make explicit the link between doubling and the 2 times table to children.

Doubling is the same as multiplying by 2.
Doubling again is the same as multiplying by 4.
Doubling *again* is the same as multiplying by 8.

If you can multiply by 3, then doubling allows you to multiply by 6.

Again, the array helps illustrate what is going on here.

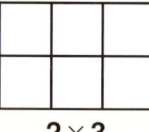

2×3

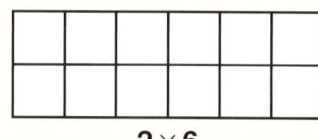

2×6

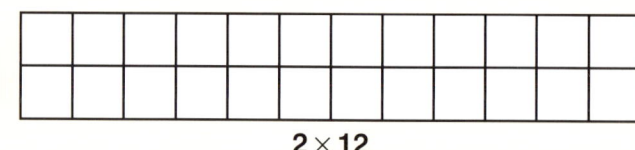

2×12

Individual and paired activities

◎ Level 1 Derive and recall multiplication facts for the 2, 5 and 10 times tables

The individual activity here provides practice in the 2, 5 and 10 times tables. Children write the answers in the grey boxes. They then find the letter that matches each answer to reveal a message.

1A Message

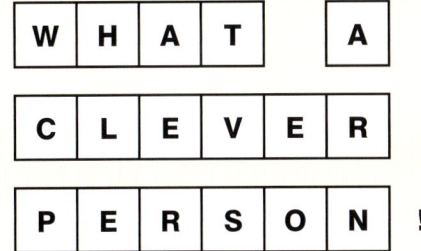

1B Message

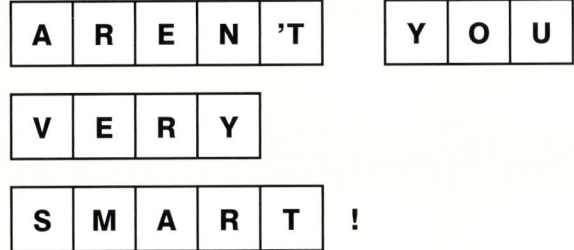

What is double three?

What is 9 times 2? Do you know that answer straight away or do you have to think about it? What do you think about?

What does your message read? Does it make sense?

In the paired activity, children work out the answers to eight 2, 5 and 10 times tables facts and colour the answers on a grid. They take turns to say their eight answers to their partner and colour their partner's eight answers on their grid. If both children are correct, they each should have coloured the same pattern.

Did you both make the same pattern? Why did you need to work together to complete the pattern?

Tell me two numbers that when multiplied together equal 30? Can you tell me another pair of numbers?

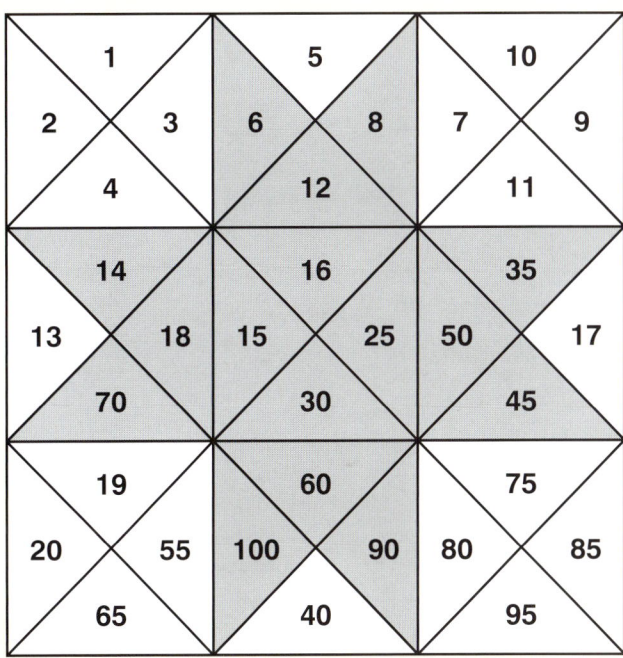

◎ Level 2 Derive and recall division facts corresponding to the 2, 5 and 10 times tables

The individual activity provides practice in the division facts corresponding to the 2, 5 and 10 times tables. Children answer 12 calculations, then compare answers with their partner. If they are correct, they should both have the same answers.

Did you and your partner get the same answer? Why did this happen?

The paired activity provides an opportunity to consolidate this still further and to practise asking and listening to division facts using a range of appropriate mathematical language.

What is 35 divided by 5? What multiplication fact did you use to help you work out this answer?

Which division facts do you know really well? Which ones do you need to work on remembering?

Further activities to develop fluency

 Point to it

Write the multiples of 2 up to the tenth multiple in random order near each other on the board.

Give two children a ruler each and ask them to stand either side of the numbers.

Ask a multiplication fact where the answer is on the board, e.g. *What is 2 times 8?*

Children point to the answer, i.e. 16.

The first child to point to the answer stays in, the other child sits down.

Another child comes out to the board.

Continue as above.

Which child can stay in the longest?

Variations

Ask all the children in the class to think of a calculation where the answer is on the board. The rest of the class asks the two children at the board questions.

Write the number 1 to 10 on the board and ask a division fact where the answer is on the board, e.g. *What is 16 divided by 2?*

 Cover the tables

Prior to the activity, decide whether you want the children to practise their 2, 5 or 10 times tables.

Provide each child with a small set of 1–10 number cards and a sheet of paper showing all the multiples to the tenth multiple of the times table being practised.

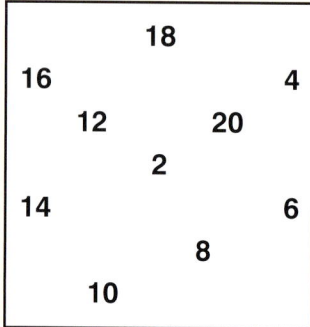
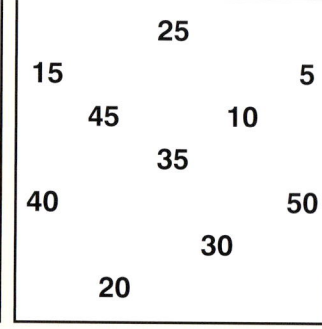
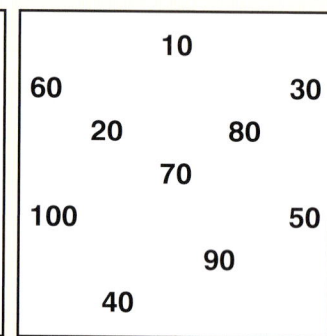

Children shuffle the cards and place them in a pile, face down. They turn over the top card – for example, 6 – say the number and multiply the number by 2, 5 or 10 (depending on which game they are playing). They then use the card to cover the answer on the sheet of paper: for example, 12 (if practising the 2 times table).

Children keep going until they have covered every number on the sheet of paper. Finally, they remove one card at a time from the paper, saying the number underneath before they do so.

Variation

Children work in pairs.

 Say it and you're out

Recite the multiples of 2 from 2 to 20 with the class. Write any two of these on the board: for example, 10, 18. Ask the children to stand behind their chairs. Go around the room reciting the multiples of 2 from 2 to 20: Child 1 says "2", Child 2 says "4", Child 3 says "6" ... (Child 11 starts with a 2 again.)

If the number called is not one of the two multiples written on the board, the child stays in; if the number called is one of the multiples on the board, the child says the number and sits down: they are out. Continue going around the class until one child is left in. That child is the winner.

Play 'Say it and you're out' for other multiplication tables.

My record sheet

Name: _____

Date: _____

	Before the activities			After the activities		
I know my 2 times table.	☺	😐	☹	☺	😐	☹
I know the divisions that go with the 2 times table.	☺	😐	☹	☺	😐	☹
I know my 5 times table.	☺	😐	☹	☺	😐	☹
I know the divisions that go with the 5 times table.	☺	😐	☹	☺	😐	☹
I know my 10 times table.	☺	😐	☹	☺	😐	☹
I know the divisions that go with the 10 times table.	☺	😐	☹	☺	😐	☹

◉ After the activities

Here are some times tables and divisions that I can answer quickly.	
Here are some times tables and divisions that are a little trickier	
Here are some ways I am going to work on this.	

Individual activity 1A

Name: _____

Date: _____

Derive and recall multiplication facts for the 2, 5 and 10 times tables

Write the answer to each of these times tables in the grey box.
Ignore the letters at the moment!

4 × 5 =	V	7 × 10 =	A
2 × 2 =	C	3 × 5 =	P
3 × 10 =	S	5 × 2 =	E
8 × 5 =	H	9 × 2 =	T
6 × 2 =	N	10 × 10 =	O
9 × 10 =	L	7 × 5 =	R

Look at the grid below.
Find the letter that matches each answer in the boxes above.
Write the letter below the number.
Some have been done for you.

40	70	18	70		4	90	10	20	10	35		15	10	35	30	100	12
W							E		E				E				!

What does it say?
If it doesn't make sense, go back and check your work.

Ask your partner your 12 questions above.
How fast can they say the answers?

Name:

Date:

Derive and recall multiplication facts for the 2, 5 and 10 times tables

Write the answer to each of these times tables in the grey box.
Ignore the letters at the moment!

8	×	2	=	[ ] N
4	×	10	=	[ ] O
3	×	2	=	[ ] A
6	×	5	=	[ ] V
7	×	10	=	[ ] T
9	×	5	=	[ ] U

2	×	5	=	[ ] R
9	×	10	=	[ ] S
5	×	5	=	[ ] Y
10	×	2	=	[ ] M
8	×	10	=	[ ] E
4	×	2	=	[ ] I

Look at the grid below.
Find the letter that matches each answer in the boxes above.
Write the letter below the number.
Some have been done for you.

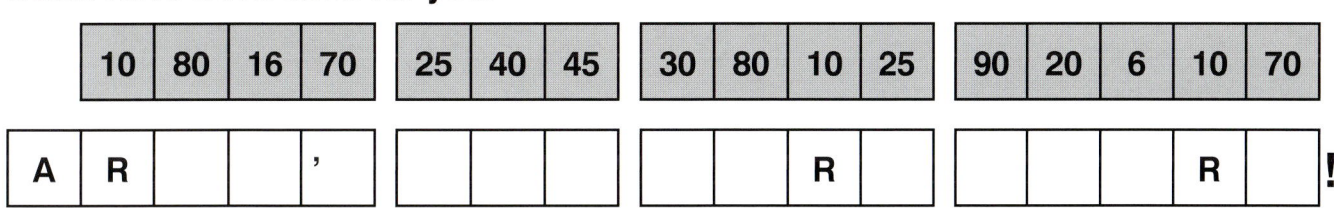

10	80	16	70		25	40	45		30	80	10	25		90	20	6	10	70
A	R			,							R						R	!

What does it say?
If it doesn't make sense, go back and check your work.

Ask your partner your 12 questions above.
How fast can they say the answers?

Paired activity 1A

Name: _____

My partner's name: _____

Date: _____

Derive and recall multiplication facts for the 2, 5 and 10 times tables

Work out the answers to these times tables facts.

You need:
- coloured pencil (same colour as your partner)

2 × 4 = △

5 × 7 = △

5 × 5 = △

2 × 7 = △

10 × 6 = △

5 × 9 = △

2 × 8 = △

10 × 10 = △

Look at your answers in the triangles △ above.

Find each of these numbers on the grid and colour them.

Speak Take turns to say your eight questions to your partner.

Listen As your partner says their eight questions to you, work out the answers and colour in the numbers on your grid.

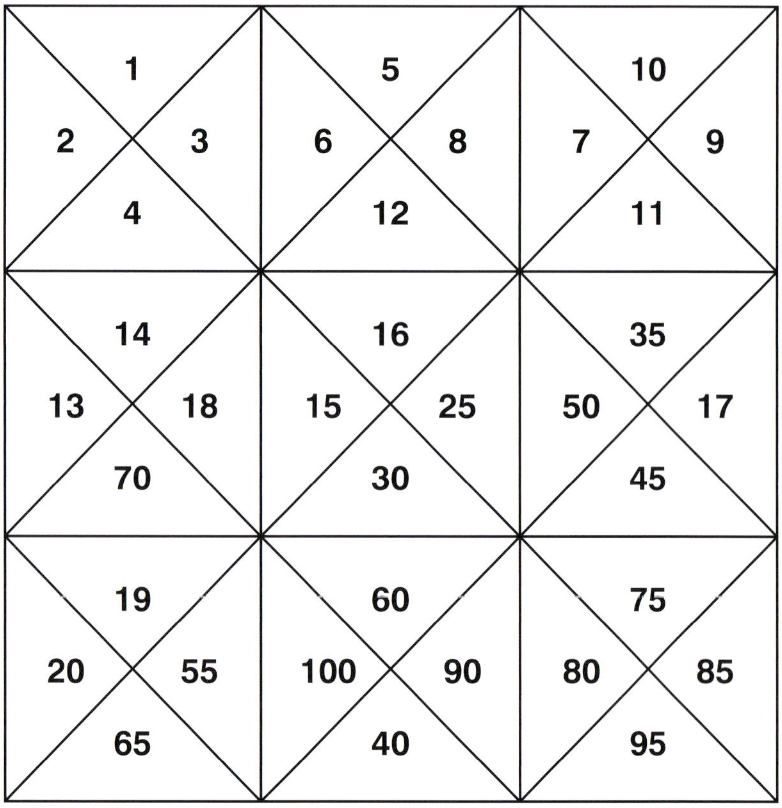

	1		5		10	
2		3	6	8	7	9
	4		12		11	
	14		16		35	
13		18	15	25	50	17
	70		30		45	
	19		60		75	
20		55	100	90	80	85
	65		40		95	

When you have both finished, compare your pictures.
Have you both drawn the same picture?

Paired activity 1B

Derive and recall multiplication facts for the 2, 5 and 10 times tables

Work out the answers to these times tables facts.

5 × 3 = △

2 × 3 = △

10 × 7 = △

5 × 6 = △

2 × 6 = △

5 × 10 = △

10 × 9 = △

2 × 9 = △

Look at your answers in the triangles △ above.

Find each of these numbers on the grid and colour them.

Speak Take turns to say your eight questions to your partner.

Listen As your partner says their eight questions to you, work out the answers and colour in the numbers on your grid.

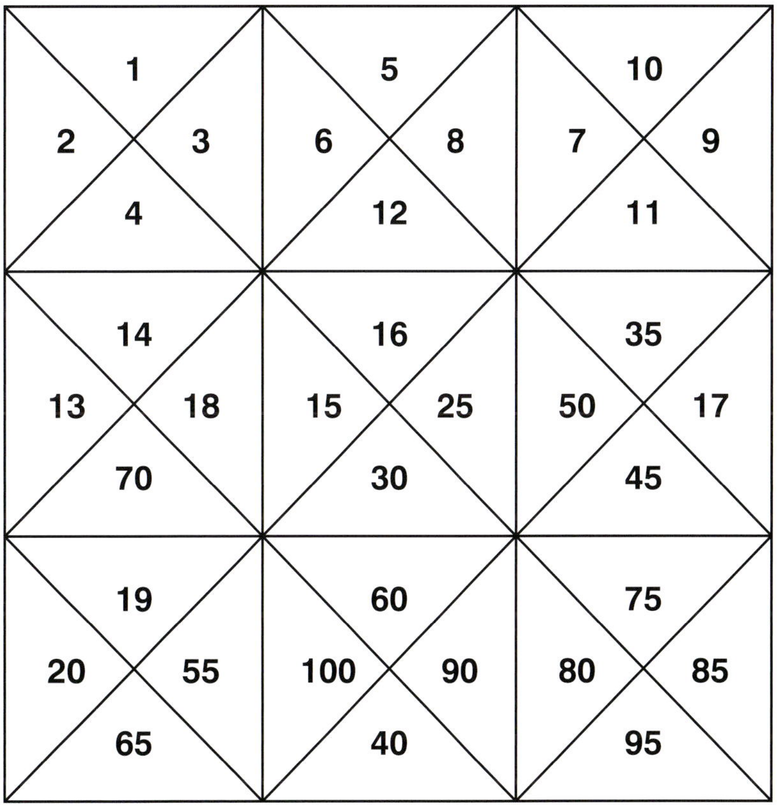

	1		5		10
2	3	6	8	7	9
	4		12		11
	14		16		35
13	18	15	25	50	17
	70		30		45
	19		60		75
20	55	100	90	80	85
	65		40		95

When you have both finished, compare your pictures. Have you both drawn the same picture?

Individual activity 2A

Derive and recall division facts corresponding to the 2, 5 and 10 times tables

Work out the answers to these division facts.

1) $15 \div 5 =$ ▢

2) $80 \div 10 =$ ▢

3) $10 \div 2 =$ ▢

4) $60 \div 10 =$ ▢

5) $16 \div 2 =$ ▢

6) $20 \div 5 =$ ▢

7) $35 \div 5 =$ ▢

8) $100 \div 10 =$ ▢

9) $20 \div 10 =$ ▢

10) $12 \div 2 =$ ▢

11) $45 \div 5 =$ ▢

12) $6 \div 2 =$ ▢

Compare your answers with your partner's.
Are they the same? If not, find out why.

Individual activity 2B

Derive and recall division facts corresponding to the 2, 5 and 10 times tables

Work out the answers to these division facts.

1) $6 \div 2 =$

2) $40 \div 5 =$

3) $50 \div 10 =$

4) $30 \div 5 =$

5) $80 \div 10 =$

6) $8 \div 2 =$

7) $70 \div 10 =$

8) $50 \div 5 =$

9) $4 \div 2 =$

10) $60 \div 10 =$

11) $18 \div 2 =$

12) $15 \div 5 =$

Compare your answers with your partner's.
Are they the same? If not, find out why.

Paired activity 2A

Name: _____

My partner's name: _____

Date: _____

Derive and recall division facts corresponding to the 2, 5 and 10 times tables

Speak Say these number sentences to your partner, but not the answers.

> **Use these words**
> divided by,
> shared equally between

1) $12 \div 2 = 6$

2) $100 \div 10 = 10$

3) $25 \div 5 = 5$

4) $40 \div 5 = 8$

5) $8 \div 2 = 4$

6) $4 \div 2 = 2$

7) $90 \div 10 = 9$

8) $35 \div 5 = 7$

9) $30 \div 10 = 3$

10) $18 \div 2 = 9$

Listen Listen to your partner and write the answers to the division number sentences they say to you.

1) ☐

2) ☐

3) ☐

4) ☐

5) ☐

6) ☐

7) ☐

8) ☐

9) ☐

10) ☐

Check your answers with your partner.

Talk with your partner about what you thought about as you worked out your answers.

Paired activity 2B

Name: _____

My partner's name: _____

Date: _____

Derive and recall division facts corresponding to the 2, 5 and 10 times tables

Listen Listen to your partner and write the answers to the division number sentences they say to you.

1)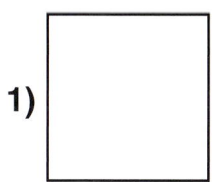

2)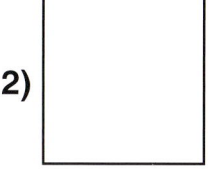

3)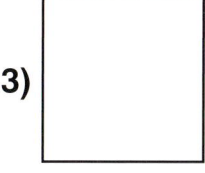

4)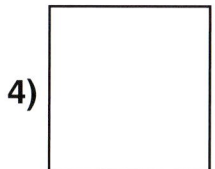

5)

6)

7)

8)

9)

10)

Speak Say these number sentences to your partner, but not the answers.

> **Use these words**
> divided by,
> shared equally between

1) $40 \div 10 = 4$

2) $30 \div 5 = 6$

3) $8 \div 2 = 4$

4) $14 \div 2 = 7$

5) $50 \div 5 = 10$

6) $80 \div 10 = 8$

7) $60 \div 10 = 6$

8) $45 \div 5 = 9$

9) $20 \div 5 = 4$

10) $16 \div 2 = 8$

Check your answers with your partner.

Talk with your partner about what you thought about as you worked out your answers.

Section 6

Mental calculation methods

Level 1
- Calculate the value of an unknown in a number sentence [for example, $30 - \square = 24$], using the symbols $+$, $-$ and $=$

Level 2
- Calculate the value of an unknown in a number sentence [for example, $\square \div 2 = 6$], using the symbols $\times$, $\div$ and $=$

Strategic approaches to develop fluency in calculating the value of an unknown in a number sentence

 ## Practise positioning of the equals (=) sign in a number sentence

Make sure that children encounter number sentences with the equals sign in a variety of positions, not just as $5+3=8$ but also $8=5+3$.

Children also need to have experiences of seeing the equals sign representing a 'balance' between two quantities: for example, $5+3=4+4$.

 ## Use correct mathematical vocabulary

It is important that children understand that the equals sign (=) is used to indicate mathematical equality. Always encourage the children to use words and phrases such as 'equals', 'is the same as' or 'balances' when referring to the equals sign. Discourage the use of the word 'makes' as this only has meaning when the calculation to be done is at the beginning of the number sentence and the 'answer' to the right.

Looking at, for example, $5+3=8$, 'five add three makes eight' is better described as 'five add three is the same as eight', because $8=5+3$ is better described as 'eight is the same as five add three' rather than 'eight equals five add three'.

 ## Practise using and applying laws of arithmetic

Children need to appreciate that from any given addition number sentence such as $5+3=8$, they can use the same three numbers to make a related addition calculation: $3+5=8$ (the commutative law). They also need to realise that they can use the same three numbers to make two subtraction facts: $8-3=5$ and $8-5=3$ (the inverse relationship).

Introducing the notion of 'number families' helps children to use and apply a known number fact to derive and recall the answer to an unknown fact. Also, if children are familiar with this idea, they will be more confident when seeing a number sentence where the value of the unknown number is in a position other than to the right of the equals sign: $5+\square=8$, $\square+3=8$, $8=5+\square$ or $8=\square+3$.

Addition and subtraction 'trios' (see Section 4) are a powerful model and image:
for example, describe the following trio as $5+\square=8$, $8=5+\square$, $8-\square=5$ or $5=8-\square$

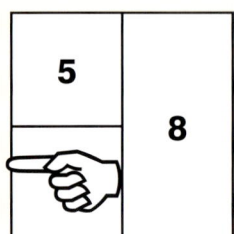

The laws of arithmetic and the 'number families' and 'trios' strategies also apply to multiplication and division.

 ## Use visual images

Practical experiences and the use of visual images help children understand what the value of an unknown number in a calculation might represent.

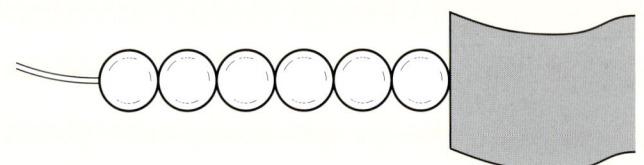

6 and how many more makes 10? ($6+\square=10$)

There are 10 beads altogether. How many are covered?

($10=6+\square$ or $10-\square=6$)

Use different symbols to represent the unknown

When presenting children with number sentences involving an unknown number or mathematical sign, make sure to use a range of different symbols.

$$\bigcirc + 3 = 8 \qquad 5 + \square = 8 \qquad 5 + 3 = \triangle \qquad 5 \diamond 3 = 8$$

Individual and paired activities

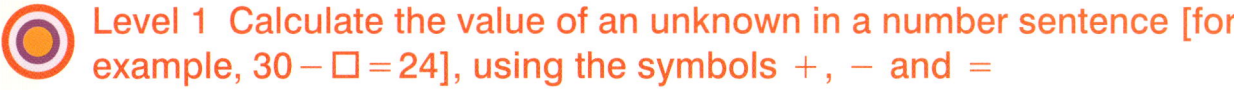 **Level 1** Calculate the value of an unknown in a number sentence [for example, $30 - \square = 24$], using the symbols $+$, $-$ and $=$

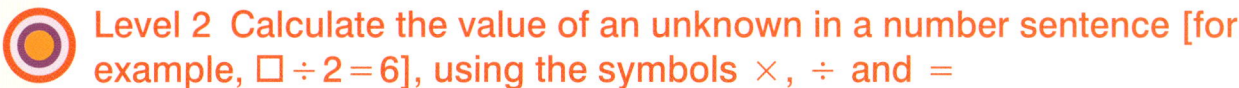 **Level 2** Calculate the value of an unknown in a number sentence [for example, $\square \div 2 = 6$], using the symbols $\times$, $\div$ and $=$

In the individual activities, children write the value of the unknown number in different shapes. They then write each of these numbers in the same shape at the bottom of their sheet. When they have finished, children compare sheets. If both children are correct, the same shapes should have the same numbers written in them.

What number did you write in the star? Did your partner have the same number written in their star?

How did you work out the missing number in this number sentence involving subtraction? Did you think about subtraction or addition? Why?

In the paired activities, children take turns to say a number. Both children use the two numbers to make a calculation and work out the answer to the calculation. If they are both correct, the children should have the same answers.

Did you both get the same answer to the first question? Did you both work it out the same way?

Which were easier to work out, the multiplication or the division number sentences? Why do you think this was?

Further activities to develop fluency

Slidey box cards

Write number sentences on a piece of card, then slide another piece of card along the sentence to show how the 'empty box' can appear in a variety of positions to represent an unknown number or symbol.

| 5 | + | | = | 8 |

| | + | 3 | = | 8 |

| 5 | + | 3 | = | |

| 5 | | 3 | = | 8 |

Do the same for multiplication and division number sentences.

| 5 | × | | = | 30 |

| | × | 6 | = | 30 |

| 5 | × | 6 | = | |

| 5 | | 6 | = | 30 |

Different possibilities

Write calculations similar to those below on the board and instruct children to do the following:

10 = □ + □ *Write numbers in the boxes to complete the number sentence.*
What other pairs of number can you use?

□ = □ + □
□ = □ − □ *Use the digits 3, 5 and 8 to complete these number sentences.*

□ + □ = □
□ − □ = □ *Write three numbers in the boxes to complete these number sentences.*

□ + □ = □ + □
□ − □ = □ − □ *Write four numbers in the boxes to complete these number sentences.*
□ − □ = □ + □

□ ○ □ = □ *Write three numbers in the boxes and choose a sign (+, −, x or ÷) to complete this number sentence.*

Do the same for multiplication and division number sentences.

Completing number sentences

Display cards showing part of a number sentence. Children suggest how they could complete the number sentence. Is there more than one solution?

| | = | 8 |

| 8 | = | |

| 5 | + | 3 | = | |

| | = | 5 | + | 3 |

Do the same for multiplication and division number sentences.

| | = | 30 |

| 30 | = | |

| 5 | × | 6 | = | |

| | = | 5 | × | 6 |

Number families

Provide each child with a sheet similar to the one on the right. Ask them to write a symbol (+, − and =) between each number to make all the number sentences correct.

Do the same for multiplication and division number sentences.

3	5	8		8	5	3
3	8	5		8	5	3
5	3	8		8	3	5
5	8	3		8	3	5

My record sheet

Name:

Date:

	Before the activities			After the activities		
I can work out the missing number in an addition number sentence: $5 + \square = 8$.	☺	😐	☹	☺	😐	☹
I can work out the missing number in a subtraction number sentence: $20 - \square = 15$.	☺	😐	☹	☺	😐	☹
I can work out the missing number in a multiplication number sentence: $4 \times \square = 20$.	☺	😐	☹	☺	😐	☹
I can work out the missing number in a division number sentence: $50 \div \square = 5$.	☺	😐	☹	☺	😐	☹

After the activities

The missing number in these addition problems is 5.	$\square + \mathbf{?} = \triangle$ $\mathbf{?} + \square = \triangle$
The missing number in these subtraction problems is 6.	$\square - \bigcirc = \mathbf{?}$ $\bigcirc - \mathbf{?} = \triangle$
The missing number in these multiplication problems is 2.	$\mathbf{?} \times \bigcirc = \triangle$ $\mathbf{?} \times \square = \triangle$
The missing number in these division problems is 6.	$\mathbf{?} \div \bigcirc = \triangle$ $\bigcirc \div \square = \mathbf{?}$

Individual activity 1A

Calculate the value of an unknown in a number sentence [for example, $30 - \square = 24$], using the symbols $+$, $-$ and $=$

Complete each number sentence.

1)   4   +   ◯   =   6

2)   5   +   ⬠   =   9

3)   7   −   ☆   =   4

4)   ◇   −   3   =   6

5)   ▭   +   2   =   10

6)   10   −   △   =   3

7)   ⯃   −   3   =   3

8)   ⌂   +   2   =   7

9)   1   +   ⬡   =   12

10)   9   −   ◯   =   8

11)   ▢   −   5   =   5

12)   ▢   +   0   =   12

Look at the answers you wrote in each of the shapes.
Write these numbers in the same shapes below.

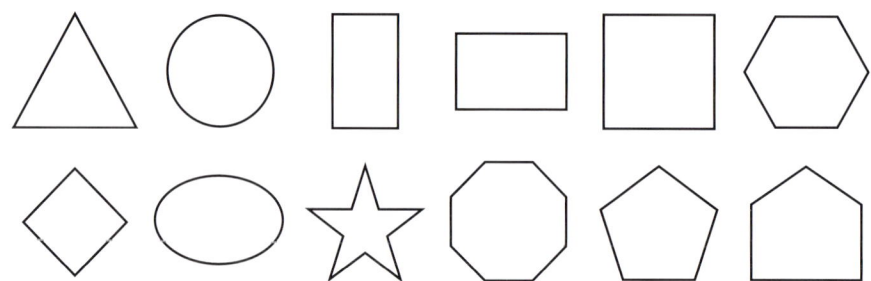

Compare these numbers with your partner.
What do you notice?

Individual activity 1B

Calculate the value of an unknown in a number sentence [for example, 30 − □ = 24], using the symbols +, − and =

Complete each number sentence.

1) ⬡ + 2 = 8

2) 5 − ◯ = 3

3) ⬡ − 1 = 10

4) 0 + □ = 12

5) 2 + ⌂ = 7

6) ▭ − 6 = 4

7) 5 − ⬭ = 4

8) ☆ + 5 = 8

9) △ − 4 = 3

10) 9 − ▯ = 1

11) 5 + ⬠ = 9

12) ◇ + 2 = 11

Look at the answers you wrote in each of the shapes.
Write these numbers in the same shapes below.

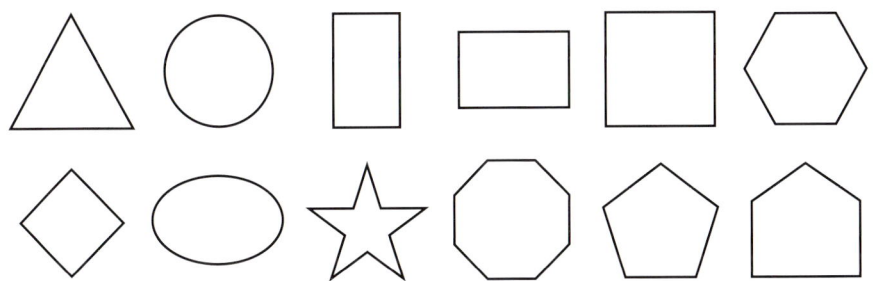

Compare these numbers with your partner.
What do you notice?

Paired activity 1A

Name: _____

My partner's name: _____

Date: _____

Calculate the value of an unknown in a number sentence [for example, $30 - \square = 24$], using the symbols $+$, $-$ and $=$

Speak Say the question number and the number in the square $\square$ to your partner, like this. *Question 1, Number 6!*

Listen Your partner will say a number back to you. Write this number in the circle $\bigcirc$ beside the same question number.

Speak **Listen** Do this until you have written a number in each circle $\bigcirc$.

1) $\boxed{6}$ $+$ $\triangle$ $=$ $\bigcirc$

2) $\triangle$ $+$ $\boxed{40}$ $=$ $\bigcirc$

3) $\triangle$ $-$ $\boxed{4}$ $=$ $\bigcirc$

4) $\boxed{15}$ $-$ $\triangle$ $=$ $\bigcirc$

5) $\triangle$ $+$ $\bigcirc$ $=$ $\boxed{37}$

6) $\triangle$ $-$ $\boxed{2}$ $=$ $\bigcirc$

7) $\boxed{24}$ $-$ $\triangle$ $=$ $\bigcirc$

8) $\bigcirc$ $+$ $\triangle$ $=$ $\boxed{67}$

9) $\triangle$ $-$ $\bigcirc$ $=$ $\boxed{3}$

10) $\boxed{20}$ $+$ $\triangle$ $=$ $\bigcirc$

11) $\triangle$ $+$ $\boxed{14}$ $=$ $\bigcirc$

12) $\bigcirc$ $-$ $\triangle$ $=$ $\boxed{50}$

Now work out the answer to each number sentence and write it in the triangle $\triangle$.

Compare your answers with your partner's and talk about any answers that are different.

Paired activity 1B

Name: _____

My partner's name: _____

Date: _____

**Calculate the value of an unknown
in a number sentence [for example, 30 − □ = 24],
using the symbols +, − and =**

Listen Listen to your partner and write the number they say
in the square □ beside the question number they tell you.

Speak For the same question number, say the number
in the circle ○ to your partner, like this. *Question 1, Number 26!*

Speak **Listen** Do this until you have written a number
in each square □.

1) □ + △ = (26)

2) △ + □ = (56)

3) △ − □ = (5)

4) □ − △ = (10)

5) △ + (33) = □

6) △ − □ = (6)

7) □ − △ = (20)

8) (40) + △ = □

9) △ − (7) = □

10) □ + △ = (31)

11) △ + □ = (64)

12) (70) − △ = □

Now work out the answer to each number sentence
and write it in the triangle △.

Compare your answers with your partner's
and talk about any answers that are different.

Individual activity 2A

Name:

Date:

Calculate the value of an unknown in a number sentence [for example, □ ÷ 2 = 6], using the symbols ×, ÷ and =

Complete each number sentence.

1) 4 × ◯ = 20

2) ⬠ ÷ 5 = 4

3) ☆ ÷ 2 = 7

4) ◇ × 2 = 16

5) ▯ × 10 = 60

6) 70 ÷ △ = 7

7) 5 × ⯃ = 35

8) ⬠(house) × 5 = 15

9) 18 ÷ ⬡ = 9

10) ⬭ ÷ 5 = 6

11) ▭ ÷ 10 = 4

12) 2 × ▢ = 18

Look at the answers you wrote in each of the shapes.
Write these numbers in the same shapes below.

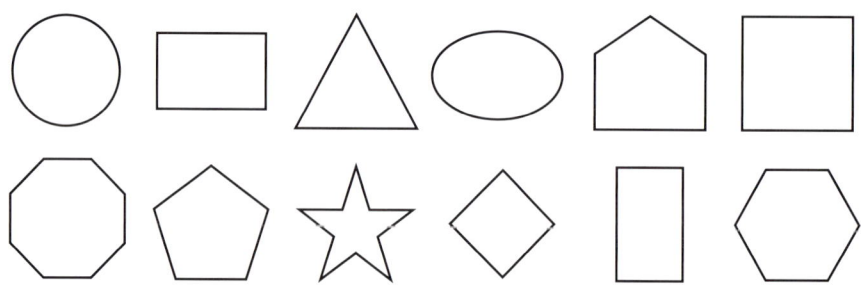

Compare these numbers with your partner.
What do you notice?

Name: _____

Date: _____

Calculate the value of an unknown in a number sentence [for example, □ ÷ 2 = 6], using the symbols ×, ÷ and =

Complete each number sentence.

1) 2 × ⬡ (octagon) = 14

2) 16 ÷ ⬡ (hexagon) = 8

3) 6 × ◯ = 30

4) ▭ ÷ 10 = 4

5) ⬭ ÷ 5 = 6

6) ▯ × 2 = 12

7) ⌂ (pentagon house) × 10 = 30

8) ☆ ÷ 2 = 7

9) 60 ÷ △ = 6

10) ◇ × 5 = 40

11) ⬠ (pentagon) ÷ 2 = 10

12) 10 × ▢ = 90

Look at the answers you wrote in each of the shapes.
Write these numbers in the same shapes below.

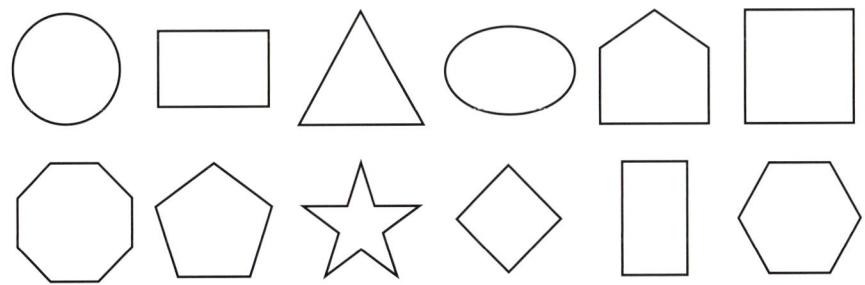

Compare these numbers with your partner.
What do you notice?

**Calculate the value of an unknown
in a number sentence [for example, $\square \div 2 = 6$],
using the symbols $\times$, $\div$ and $=$**

Speak Say the question number and the number in the square $\square$
to your partner, like this. *Question 1, Number 6!*

Listen Your partner will say a number back to you.
Write this number in the circle $\bigcirc$ beside the same question number.

Speak **Listen** Do this until you have written a number
in each circle $\bigcirc$.

1) $\triangle \times \boxed{6} = \bigcirc$

2) $\bigcirc \div \triangle = \boxed{10}$

3) $\boxed{5} \times \triangle = \bigcirc$

4) $\triangle \times \boxed{2} = \bigcirc$

5) $\triangle \times \bigcirc = \boxed{40}$

6) $\triangle \div \boxed{5} = \bigcirc$

7) $\triangle \div \boxed{2} = \bigcirc$

8) $\boxed{90} \div \triangle = \bigcirc$

9) $\triangle \div \bigcirc = \boxed{6}$

10) $\boxed{5} \times \triangle = \bigcirc$

11) $\boxed{45} \div \triangle = \bigcirc$

12) $\bigcirc \times \triangle = \boxed{30}$

Work out the answer to each number sentence
and write it in the triangle $\triangle$.

Compare your answers with your partner's
and talk about any answers that are different.

Paired activity 2B

Name:

My partner's name:

Date:

**Calculate the value of an unknown
in a number sentence [for example, □ − 2 = 6],
using the symbols ×, ÷ and =**

Listen Listen to your partner and write the number they say
in the square ☐ beside the question number they tell you.

Speak For the same question number, say the number
in the circle ◯ to your partner, like this. *Question 1, Number 12!*

Speak **Listen** Do this until you have written a number
in each square ☐.

1) △ × ☐ = ⬤12   7) △ ÷ ☐ = ⬤7

2) ⬤20 ÷ △ = ☐   8) ☐ ÷ △ = ⬤9

3) ☐ × △ = ⬤35   9) △ ÷ ⬤10 = ☐

4) △ × ☐ = ⬤16   10) ☐ × △ = ⬤25

5) △ × ⬤10 = ☐   11) ☐ ÷ △ = ⬤9

6) △ ÷ ☐ = ⬤7   12) ⬤3 × △ = ☐

Work out the answer to each number sentence
and write it in the triangle △.

Compare your answers with your partner's
and talk about any answers that are different.